Applied Nuclear Physics Conference (ANPC 2021)

EPJ Web of Conferences Volume 261 (2022)

Prague, Czech Republic
12 – 17 September 2021

Editors:

Anna Mackova
Katharina Lorenz
Andre Vantomme

ISBN: 978-1-7138-4832-5

Printed from e-media with permission by:

Curran Associates, Inc.
57 Morehouse Lane
Red Hook, NY 12571

Some format issues inherent in the e-media version may also appear in this print version.

For additional information, please contact EDP Sciences – Web of Conferences
at the address below.

EDP Sciences – Web of Conferences
17, Avenue du Hoggar
Parc d'Activité de Courtabœuf
BP 112
F-91944 Les Ulis Cedex A
France

Phone: +33 (0) 1 69 18 75 75
Fax: +33 (0) 1 69 28 84 91

contact-edps@webofconferences.org

Additional copies of this publication are available from:

Curran Associates, Inc.
57 Morehouse Lane
Red Hook, NY 12571 USA
Phone: 845-758-0400
Fax: 845-758-2633
Email: curran@proceedings.com
Web: www.proceedings.com

TABLE OF CONTENTS

Editorial

Applied Nuclear Physics Conference 2021
September 12–16
Prague, Czech Republic

The first Applied Nuclear Physics Conference (ANPC 2021) was held in Prague, Czech Republic, 12-16 September 2021. The European Physical Society (EPS) introduced the new Applied Nuclear Physics Conference promoted by the Nuclear Physics Division (NPD) of EPS. The objective of the NPD EPS is to contribute to the advancement of nuclear physics in Europe and neighbouring countries by providing a forum for networking of scientists, discussion of scientific subjects, and supporting the knowledge dissemination in the field of nuclear sciences via common events like, e.g., conferences.

The ANPC conference provided an open forum for scientists and engineers working in the wide field of nuclear applications with particular emphasis on energy, health, space, environment, materials science, and preservation of cultural heritage. This was a unique opportunity for scientists from these different communities to exchange views and foster synergies to find new ways and horizons to the challenges in their fields. Moreover, the ANPC offered an overview of the current state-of-the-art to young researchers and students, via numerous lectures presented by leading scientists in a broad range of nuclear physics applications. The hybrid conference, organized by the Nuclear Physics Institute of the Czech Academy of Sciences (NPI), Czech Republic, was chaired by Anna Mackova (NPI) and co-chaired by Jan Kucera (NPI) and Jan John (Faculty of Nuclear Science and Physical Engineering, Czech Technical University, Prague, Czech Republic). ANPC hosted 124 registered participants, 66 onsite and 58 online.

We are very proud that ANPC hosted two IBA-Europhysics prize winners as plenary and invited speaker, Marco Durante (GSI, Biophysics Department and Technische Universität Darmstadt, Germany, awardee 2013) and Thomas Haberer (Heidelberg University Hospital, Germany, awardee 2020). ANPC was honoured by the opportunity to also host the IBA-

Europhysics prize ceremony, where Thomas Haberer received the prize from the hands of the NPD EPS member Iva Bogdanovic Radovic (Ruder-Boskovic Institute, Croatia).

We are very grateful that ANPC was carefully managed and supervised by the International Advisory Board including leading scientist in nuclear physics from Europe as well as non-European countries which recommended and promoted the invited and plenary speakers. The scientific program has been put together with great responsibility by the International Program Committee. ANPC offered a variety of excellent speakers from the ion beam physics community working in material science, nuclear physics applications in medicine, environmental and cultural heritage studies. ANPC provided a total of 5 plenary talks, 26 invited talks, 56 contributed talks and 35 poster presentations.

We would like to thank all sponsors and exhibitors to make our event attractive and fruitful. Particularly, we thank the following scientific institutions and companies: iThemba LABS (Laboratory for Accelerator Based Sciences), Center of Accelerators and Nuclear Analytical Methods of NPI, National Electrostatics Corporation, Vakuum Servis, Vakuum Praha, Pfeiffer Vacuum, Specion and Jan Evangelista Purkyne University in Usti nad Labem. MDPI Publishing supported and promoted the ANPC conference and the Best Poster Award.

The ANPC Proceedings are a collection of the contributed papers presented in the following sections

- Ion Beam Analytical Methods in Material Science,
- Ion and Neutron Beam Irradiation of Materials,
- Nuclear Physics for Energy and Space Technologies,
- Nuclear Analytical Methods for Environmental and Cultural Heritage Studies,
- Nuclear Physics in Medicine.

We believe that ANPC proved to be a very attractive platform for interdisciplinary scientific discussions and will open new synergies between various nuclear physics applications. It was proposed to organize the ANPC conference every 3 years. The place and chair for hosting the next ANPC will be selected by the EPS NPD soon. Sincerely

Anna Macková

Katharina Lorenz

André Vantomme

ANPC Proceedings Editors

edp sciences

Web of Conferences

Statement of Peer review

In submitting conference proceedings to *Web of Conferences*, the editors of the proceedings certify to the Publisher that

1. They adhere to its **Policy on Publishing Integrity** in order to safeguard good scientific practice in publishing.
2. All articles have been subjected to peer review administered by the proceedings editors.
3. Reviews have been conducted by expert referees, who have been requested to provide unbiased and constructive comments aimed, whenever possible, at improving the work.
4. Proceedings editors have taken all reasonable steps to ensure the quality of the materials they publish and their decision to accept or reject a paper for publication has been based only on the merits of the work and the relevance to the journal.

Title, date and place of the conference

APPLIED NUCLEAR PHYSICS CONFERENCE (ANPC)
September 12-17, 2021
Prague, Czech Republic

Proceedings editor(s):

Anna Mackova (NPI CAS, Czech Republic)
Katharina Lorenz (University of Lisbon and INESC MN, Portugal)
Andre Vantomme (KU Leuven, Belgium)

Date and editor's signature

22. 3. 2022 Macková Anna

Assinado por: **Katharina Lorenz**
Num. de Identificação: 32304993
Data: 2022.03.22 12:28:36+00'00'

André Vantomme (Signature)

Digitally signed by André Vantomme (Signature)
Date: 2022.03.22 14:25:42 +01'00'

Oxygen mobility in yttrium hydride films studied by isotopic labelling

Dmitrii Moldarev[1,2*], *Camille* Aracheloff[1], *Marcos V.* Moro[1], *Eduardo* Pitthan[1], *Max* Wolff[1] and *Daniel* Primetzhofer[1]

[1]Department of Physics and Astronomy, Uppsala University, Box 516, 751 20 Uppsala, Sweden
[2]Department of Material Science, Moscow Engineering Physics Institute, 115409, Moscow, Russia

Abstract. The photochromic properties of oxygen-containing yttrium hydride thin films are directly dependent on the oxygen concentration in the material. We use $^{16}O/^{18}O$ labelling to study oxidation of YH_2 films. Oxygen penetrates the film through grain boundaries and intercolumnar voids oxidising the whole film thickness, without pronounced surface oxidation or self-passivation. Once oxidised, the mobility of oxygen in the films is low and no detectable changes in chemical composition of ^{18}O-labeled YHO films is found under illumination.

1 Introduction

The nearly exponential rise of the global energy demand over the last decades necessitates the development of sustainable technologies [1]. Smart windows based on photochromic materials represent a promising technology to save energy as they enable passive control of heat and light flow between buildings and the environment. Oxygen-containing yttrium hydride (YHO) [2] and some other oxygen-containing rare-earth metal hydrides (REMHO) [3], [4] in general, may show reversible photochromism – a modulation of the optical transmittance under illumination. The effect is observed at room temperature and ambient conditions and accompanied by a persistent photoconductivity [5]. Unlike their transition metal oxide counterparts, photodarkening in REMHO happens uniformly over the visible range without pronounced absorption bands [6], [7]. In practice, it implies that films do not change their color but transparency, which makes this material a good candidate for application in smart windows.

Photochromic YHO thin films are commonly produced in a two-step process. In the first step, a thin film of yttrium dihydride (YH_2) is deposited on a substrate using either reactive magnetron sputtering [8] or e^--beam evaporation [9]. After deposition the film is exposed to oxygen, often by venting the sputtering chamber, and oxidized. Thus, the opaque metal hydride turns into a transparent and photochromic oxygen containing metal hydride [10]. The amount of oxygen incorporated into the film at the final step determines the photochromic behavior of YHO. Photochromism is found for oxygen to yttrium ratios between 0.5 and 1.5, with the oxygen replacing hydrogen atoms under oxidation. Moreover, a general trend of decreasing photochromic activity with increasing oxygen concentration is reported [3], [11].

* Corresponding author: dmitry.moldarev@physics.uu.se

A comprehensive understanding of the oxidation process of YH_2 is vital for controlled synthesis of photochromic films with tailored properties.

Despite the high relevance of YHO for potential technological applications, the nature of its photochromic effect remains under debate. In Ref. [12], a reversible lattice contraction under illumination was reported, suggesting atomic rearrangements being related to the photodarkening. Positron annihilation and muon spin relaxation experiments suggest anion transport in connection with the photochromic effect [13], [14]. Along this line Baba et al. propose a light-induced oxygen breathing during photodarkening and bleaching [15]. However, experiments in high vacuum [16], [17] and with films encapsulated by diffusion barriers [18] show a persistent photochromic effect without the need of material exchange with the environment. Material transport, oxygen or hydrogen, within the film may still be possible. Recent findings suggest that photochromic films are composed of two phases [19] with a hydrogen rich columnar structure separated by oxygen rich regions. Material transport between these two phases may have the potential to explain the photochromism.

In this work, we study the oxidation of YH_2 and oxygen mobility in YHO. We use bilayer structures of YH_2/YHO and show that oxygen does not migrate from the oxidized layer into the dihydride. To trace the oxygen mobility and exchange with the environment we partially oxidize films in $^{18}O_2$ enriched atmosphere and study the redistribution under exposure to air. Finally, we investigate light-induced changes in chemical composition of YHO films and find no long-range oxygen diffusion.

2 Experimental details

2.1 *In-situ* experiment

The experiments were conducted in the scattering chamber SIGMA at the 5 MV 15-SDH-2 tandem accelerator (Tandem laboratory) at Uppsala University (Sweden). The setup is equipped with Passivated Implanted Planar Silicon detectors (PIPS), a Silicon Drift Detector (SDD) and an e^--beam evaporator (FOCUS EFM 3T), enabling *in-situ* ion beam analysis (IBA) of thin films under deposition. A leak valve installed between the main chamber and a gas feed system allows different gas environments, whereas a large optical viewport is used to illuminate samples inside the chamber with light. For more details of the scattering chamber, we refer to literature [20].

A bilayer film consisting of an YH_2 layer on top of YHO was deposited onto a C substrate. The choice of the substrate allows the unambiguous identification of O in the thin films. Metallic Y (nominal purity: 99%) was evaporated in H_2 atmosphere ($P=3\times10^{-6}$ mbar) for 13 min. The resulting YH_2 layer turned photochromic upon exposure to air forming YHO. Films deposited at the same deposition conditions yielded photochromic films previously [9]. After oxidation, another YH_2 layer was grown onto the YHO film under identical conditions. The bilayer structure (YH_2/YHO) with a gradient in the anion concentration enables to study diffusion and anion mobility. For photodarkening we used a blue LED-array (λ=455 nm, $W_{flux}\approx$500 $mWcm^{-2}$) as employed in Ref. [9].

The chemical composition and depth distribution of elements and their isotopes were determined by Rutherford Backscattering Spectrometry (RBS) using 2 MeV $^4He^+$ after each synthesis step. The angle between the incident ions and the PIPS detector (i.e. scattering angle) was 170°. The normal of the sample surface was rotated 140° with respect to the direct beam. This geometry improves the depth resolution due to an increased path length of the probing ions in the film. Characteristic x-rays emitted under He irradiation were detected by the SDD. L_α x-rays of Y were used to normalize the RBS spectra. All RBS spectra were analyzed using SIMNRA [21].

2.2 *Ex-situ* analysis

After the *in-situ* experiments the samples were studied *ex-situ*. Time of Flight – Energy Elastic Recoil Detection Analysis (ToF-E ERDA) was used to quantify the depth distribution of all constituents. The ^{18}O content in the films was extracted from Nuclear Reaction Analysis (NRA) with enhanced depth resolution.

ToF-E ERDA measurements were carried out at the tandem accelerator using 36 MeV $^{127}I^{8+}$ as probing beam. Recoil species from the target were detected at a detection angle of 45° by a combined segmented anode gas ionization chamber and time-of-flight detector. More technical details on ToF-E can be found elsewhere [22]. The incident angle was 67.5°. Possible sources of systematic uncertainties of ERDA measurements were discussed in detail in Ref. [17]. Depth profiles of chemical elements were derived using Potku [23].

Profiling of ^{18}O was done at the single stage 350 kV Danfysik implanter (Tandem Laboratory, Uppsala University) using the $^{18}O(p, \alpha)^{15}N$ nuclear reaction and taking advantage of its sharp resonance at 151 keV (Γ_R = 100 eV). A depth profile is obtained by increasing the energy of incident protons stepwise starting from energies below the resonance. Alpha particles originating from the nuclear reaction are detected by a PIPS detector with large solid angle placed at 150° mean detection angle and covered by a 4 μm Al foil to avoid backscattered particles. The sample normal was 10° from the incident beam direction. For more details on instrumentation employed for NRA measurements we refer to Ref. [24].

3 Results and discussions

RBS measurements on the bilayer sample were conducted immediately after deposition and repeated 30 minutes later. In between the measurements, the sample was stored under high vacuum ($P \approx 5\times10^{-8}$ mbar) conditions. Figure 1a) shows both RBS spectra normalized. The data were normalized using the substrate signal. Figure 1b) is a zoom into the energy region of the O signal. The vertical dashed grey line indicates the interface between the oxidised layer and the additional layer after deposition but before oxidation. The films (black open circles in Figure 1) contain uniformly distributed O at a level of 36 at. % in the bottom layer and up to 5 at. % of O in the top one. The oxygen impurity may be related to residual gas in the pipes used to inject H_2 during evaporation. The distribution of all chemical elements remains the same after 30 min (red crosses in Figure 1) apart from an O-enriched layer appearing right at the surface. A fit to the O peak at the surface yields 1.4×10^{16} at/cm^2 added oxygen atoms.

At a pressure of 5×10^{-8} mbar and a temperature of 22°C the particle flux of air is 1.42×10^{13} cm^{-2}s^{-1} [25]. The number of particles hitting the surface of the sample in 30 minutes is 2.5×10^{16} at/cm^2, which is in good agreement with the number extracted from the RBS measurement, assuming that most of the residual gas is oxygen, water or hydroxyl. We conclude that the residual gas in the chamber is the reason for the observed surface oxidation but no material transport between the layers in the film is detected.

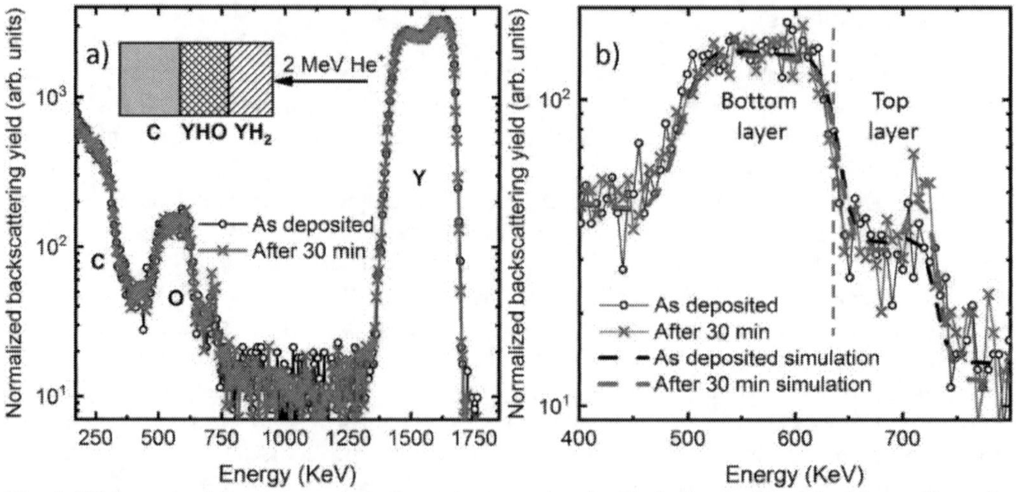

Fig. 1. RBS spectra (a) and zoom-in in the region associated with O signals (b) of the bilayer film taken as deposited (black open circles) and after 30 min in HV (red crosses). Black and red dashed lines show fit done using SIMNRA. The vertical dashed grey line indicates the interface between the two layers.

To study the oxidation process in more detail a second bilayer sample (YH_2/YHO) was grown but this time oxidized in $^{18}O_2$ (isotopically enriched to 97%) at a pressure of 3×10^{-3} mbar for 3 min and subsequently exposed to air. Figure 2a) shows the part of RBS spectra with the O-signal of the as-deposited film (black-open circles) and after oxidation in $^{18}O_2$ (red crosses) as well as after exposure to air (blue squares). The vertical dashed grey line indicates the interface between the two layers. Exposure to $^{18}O_2$ results in partial oxidation of the top layer with both ^{16}O and ^{18}O. The presence of ^{16}O is attributed to contamination of the gas feed system. Further exposure to air leads to continuous oxidation. Remarkably, the additional oxygen, which is predominantly ^{16}O does not replace ^{18}O nor redistributes it but rather further oxidizes the film throughout its full thickness. This fact supports the hypothesis that oxidation progresses along permeable grain boundaries and intercolumnar voids. Moreover, the result clearly shows that oxygen atoms are tightly bound once the oxide is formed. These results provide further evidence for the hypothesis of a two phase nature of the material with a columnar structure as found in GdHO films and revealed by transmission electron microscopy [19].

Films deposited at identical conditions on CaF_2 substrates show photochromic properties [9]. Therefore, it is plausible to assume that the films studied in this work are also transparent and photochromic, despite the opaque nature of the substrate obscuring the optical properties but allowing a detailed study of the oxygen inventory with RBS. A potential influence of illumination on material migration, oxygen diffusion, was investigated. RBS spectra recorded before and after 10 min of illumination and shown as black open circles and red crosses, respectively, are presented in Figure 2b). No detectable changes in chemical composition are observed, excluding light-induced oxygen transport over long distances but leaving a possibility of a local diffusion over distance smaller than the depth resolution, which is several 10's of nm.

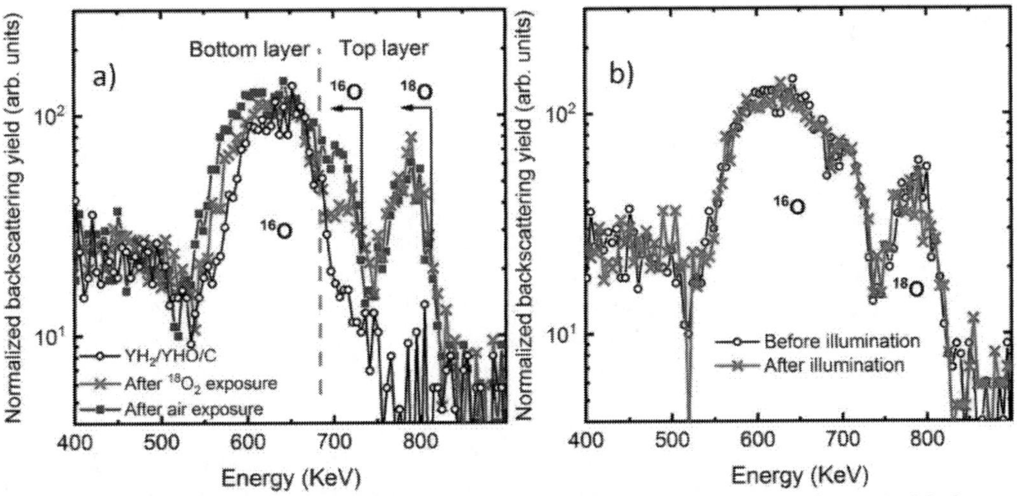

Fig. 2. Oxygen RBS signals of the bilayer film (a) recorded for the as deposited sample (black open circles), after exposure to $^{18}O_2$ of pressure 3×10^{-3} mbar for 3 min (red crosses) and oxidation in air (blue squares). Oxygen signals (b) before (black open circles) and after 10 min of illumination (red crosses).

The sample studied under illumination was characterized by ToF-E ERDA and NRA 4 days and 13 days after deposition. Figure 3a) shows that all chemical elements including O are uniformly distributed throughout the film despite of the fact that the sample was grown in two layers. The presence of C in the film (8-9 at. %) is explained by discontinuities in the film due to partial delamination over time. It is known that the stability of YHO films in air and produced by e$^-$-beam evaporation is lower as compared to the ones produced by magnetron sputtering [9]. Both ERDA and NRA (Figure 3a and 3b, respectively) confirm that ^{18}O is primarily concentrated in the top layer again suggesting progressive oxidation along grain boundaries not affecting existing oxidic domains.

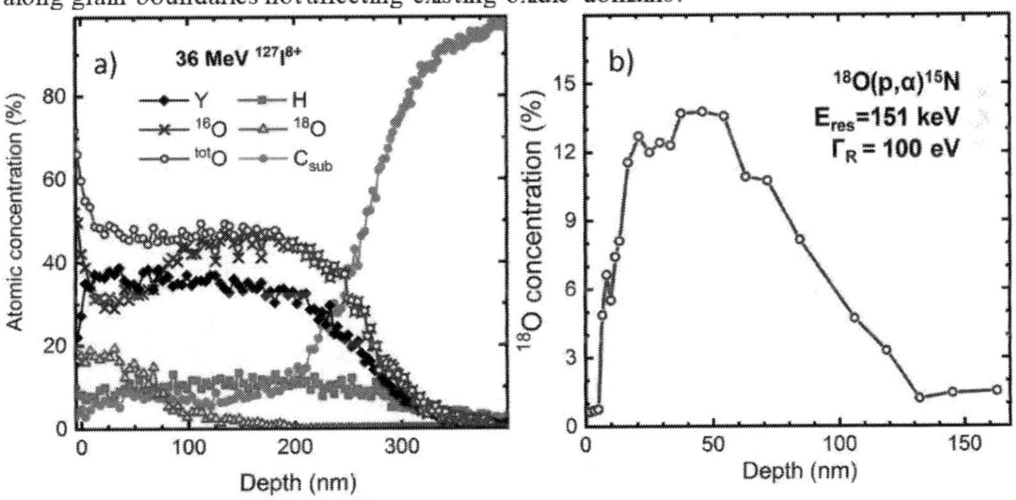

Fig. 3. Depth profile of chemical elements for the fully oxidized bilayer YHO film derived from ToF-E ERDA (a). Depth distribution of ^{18}O in the YHO film calculated from NRA measurements (b).

4 Summary and conclusions

We have prepared bilayer YH_2/YHO films and studied their oxidation and oxygen mobility by tracking the chemical composition using isotope-sensitive IBA. We show that a gradient in the anion concentration, artificially created in the film, persists over time, demonstrating low oxygen mobility. Using ^{18}O as a tracer, we conclude that oxidation of YH_2 progresses through grain boundaries and intercolumnar voids not effecting pre-existing O. The result implies that the oxidation rate may be controlled via the films microstructure. Composition analysis of isotope-labeled photochromic YHO film before and after illumination reveals no detectable changes. This proves that long-range diffusion of anions can not be responsible for photodarkening and supports the hypothesis of inter-phase H transfer, as proposed in Ref. [19].

The authors would like to thank the Swedish Research Council VR-RFI (contracts No. 2017-00646-9 and 2019-00191) and the Swedish Foundation for Strategic Research (contract RIF14-0053) for supporting accelerator operation. DM thanks the Swedish Institute for the stipend for his research visit at Uppsala University granted as part of the Visby program. CA acknowledges the ERASMUS+ Programme which supported her stay at Uppsala University.

References

[1] T. Ahmad and D. Zhang, Energy Reports, **6**, 1973–1991 (2020).

[2] T. Mongstad, C. Platzer-Björkman, J.P. Maehlen, L.P.A. Mooij, Y. Pivak, B. Dam, E.S. Marstein, B.C. Hauback, and S.Z. Karazhanov, Sol. Energy Mater. Sol. Cells, **95**, no. 12, 3596–3599 (2011).

[3] S. M. M. Aðalsteinsson, M. V. V Moro, D. Moldarev, S. Droulias, M. Wolff, and D. Primetzhofer, Nucl. Instruments Methods Phys. Res. Sect. B Beam Interact. with Mater. Atoms, **485**, 36–40 (2020).

[4] F. Nafezarefi, H. Schreuders, B. Dam, and S. Cornelius, Appl. Phys. Lett., **111**, no. 10, 103903 (2017).

[5] C. C. You, T. Mongstad, E. S. Marstein, and S. Z. Karazhanov, Materialia, **6**, no. April, 100307 (2019).

[6] T. He and J. N. Yao, Res. Chem. Intermed., **30**, no. 4–5, 459–488 (2004).

[7] A. B. A. Kayani S. Kuriakose, M. Monshipouri, F.A. Khalid, S. Walia, S. Sriram, and M. Bhaskaran, Small, **17**, no. 32, 2100621 (2021).

[8] J. Montero, F. A. Martinsen, M. Lelis, S. Z. Karazhanov, B. C. Hauback, and E. S. Marstein, Sol. Energy Mater. Sol. Cells, **177**, 106–109 (2018).

[9] K. Kantre, M. V. Moro, D. Moldarev, M. Wolff, and D. Primetzhofer, Scripta Materialia, **186**. 352–356 (2020).

[10] C. C. You, D. Moldarev, T. Mongstad, D. Primetzhofer, M. Wolff, E.S. Marstein, and S.Z. Karazhanov, Sol. Energy Mater. Sol. Cells, **166** (2017).

[11] D. Moldarev, M. V Moro, C.C. You, E.M. Baba, S.Z. Karazhanov, M. Wolff, and D. Primetzhofer, Phys. Rev. Mater., **2**, no. 11, 115203 (2018).

[12] J. P. Maehlen, T. T. Mongstad, C. C. You, and S. Karazhanov, J. Alloys Compd., **580**, S119–S121 (2013).

[13] M. P. Plokker, S.W.H. Eijt, F. Naziris, H. Schut, F. Nafezarefi, H. Schreuders, S. Cornelius, and B. Dam, Sol. Energy Mater. Sol. Cells, **177**, no. March 2017, 97–

105 (2018).

[14] D. Chaykina, T. de Krom, G. Colombi, H. Schreuders, A. Suter, T. Prokscha, B. Dam, and S. Eijt, Phys. Rev. B, **103**, no. 22, 224106 (2021).

[15] E. M. Baba, J. Montero, E. Strugovshchikov, E. Ö. Zayim, and S. Karazhanov, Phys. Rev. Mater., **4**, no. 2, 1–8 (2020).

[16] D. Moldarev, L. Stolz, M. V Moro, S.M. Aðalsteinsson, I.-A. Chioar, S.Z. Karazhanov, D. Primetzhofer, and M. Wolff, J. Appl. Phys., **129**, no. 15, 153101 (2021).

[17] M. V Moro, D. Moldarev, C.C. You, E.M. Baba, S.Z. Karazhanov, M. Wolff, and D. Primetzhofer, Sol. Energy Mater. Sol. Cells, **201**, 110119 (2019).

[18] M. V Moro, S.M. Aðalsteinsson, T.T. Tran, D. Moldarev, A. Samanta, M. Wolff, and D. Primetzhofer, Phys. status solidi – Rapid Res. Lett., vol. 15, no. 6, p. 2000608, Jun. 2021.

[19] M. Hans, T.T. Tran, S.M. Aðalsteinsson, D. Moldarev, M. V. Moro, M. Wolff, and D. Primetzhofer, Adv. Opt. Mater., **2000822**, 1–6 (2020).

[20] K. Kantre, M.V. Moro, D. Moldarev, D. Johansson, D. Wessman, M. Wolff, and D. Primetzhofer, Nucl. Instruments Methods Phys. Res. Sect. B Beam Interact. with Mater. Atoms, **463** (2020).

[21] M. Mayer, W. Eckstein, H. Langhuth, F. Schiettekatte, and U. Von Toussaint, Nucl. Instruments Methods Phys. Res. Sect. B Beam Interact. with Mater. Atoms, **269**, no. 24, 3006–3013 (2011).

[22] P. Ström, P. Petersson, M. Rubel, and G. Possnert, Rev. Sci. Instrum., **87**, no. 10, 103303 (2016).

[23] K. Arstila, J. Julin, M.I. Laitinen, J. Aalto, T. Konu, S. Kärkkäinen, S. Rahkonen, M. Raunio, J. Itkonen, J.-P. Santanen, T. Tuovinen, and T. Sajavaara, Nucl. Instruments Methods Phys. Res. Sect. B Beam Interact. with Mater. Atoms, **331**, 34–41 (2014).

[24] S. A. Corrêa, E. Pitthan, M. V Moro, and D. Primetzhofer, Nucl. Instruments Methods Phys. Res. Sect. B Beam Interact. with Mater. Atoms, **478**, 104–110 (2020).

[25] J. F. O'Hanlon, "Gas Properties," A User's Guide to Vacuum Technology, 9–24 (2003).

Recent developments in IBA analysis at CENTA, Bratislava

Miroslav Ješkovský[1*], *Jakub* Kaizer[1], *Ivan* Kontuľ[1], *Jakub* Kvasniak[1], *Ján* Pánik[2], *Jakub* Zeman[1] and *Pavel P.* Povinec[1]

[1]Centre for Nuclear and Accelerator Technologies (CENTA), Faculty of Mathematics, Physics and Informatics, Comenius University, 842 48 Bratislava, Slovakia
[2]Institute for Medical Physics, Biophysics, Informatics and Telemedicine, Faculty of Medicine, Comenius University, 813 72, Bratislava, Slovakia

Abstract. An experimental setup used at the CENTA (Centre for Nuclear and Accelerator Technologies) laboratory to carry out IBA (Ion Beam Analysis) techniques is described. PIXE (Particle Induced X-ray Emission) technique offers a unique way of non-destructive elemental analysis using accelerated ion beams. The calibration of instrumental constants (H-values) of SDD and BEGe detectors, used in the setup, is described in detail. The H-value was determined as a function of X-ray energy by using thin MicroMatter standards of chosen elements. The calibrated H-values were then used in GUPIXWIN for the determination of elemental concentrations in two IAEA reference materials, which were compared with reference values. A reasonable agreement was achieved between the measured and reference values, while the calibrated H-values will be used for further analyses.

1 Introduction

Particle induced X-ray emission is a non-destructive analytical technique used for studying surface elemental composition, with high sensitivity for multiple elements. It is frequently applied as a complementary technique to other IBA techniques, such as Rutherford backscattering spectrometry (RBS), nuclear reaction analysis (NRA), and elastic recoil detection analysis (ERDA), to obtain more complex information about the analysed sample. Possibilities of PIXE techniques to determine the elemental composition of thin and thick samples were demonstrated in many applications, e.g. [1-7]. In the case of thin films, the intensity of measured X-rays depends on the accumulated beam charge, the ionisation cross-section, the fluorescence yield, the transmission through filters (present between a sample and the detector), and the detector efficiency [8]. For the thick samples, also stopping power of the ion beam, and attenuation of the X-rays in the sample complicate the calculation of elemental concentration [9].

* Corresponding author: miroslav.jeskovsky@uniba.sk

The calibration of the PIXE system is needed when quantitative analysis is involved. The instrumental constant H-value was proposed to include the geometry of the setup [10], and it has been frequently used for the determination of elemental concentrations [11-13].

2 Experimental setup

CENTA laboratory was established in 2013 at the Faculty of Mathematics, Physics and Informatics of the Comenius University. The ion beam facility has a 3 MV Pelletron tandem accelerator dedicated to IBA techniques and the accelerator mass spectrometry (scheduled for 2022). A detailed description of the laboratory and its accelerator system can be found in [14]. IAEA funded the installation of the PIXE/PIGE (Particle Induced Gamma-ray Emission) beamline at the CENTA in 2015 [15], and a new detection system was incorporated into the chamber used for IBA analysis in 2019. The current experimental setup allows simultaneous PIXE, PIGE and RBS analysis.

2.1 IBA detection system

The detection system includes three separate detectors. Silicon drifted detector of 70 mm^2 FAST SDD (Amptek) in X-123 configuration with C2 (Si_3N_4) entrance window is used for PIXE analysis of light elements (from carbon to yttrium). It is suitable for X-ray detection in the range of 0.1-15 keV with an energy resolution of 122 eV for 5.9 keV (^{55}Fe). Germanium BE2825 BEGe detector (Mirion, former Canberra) with 0.6 mm carbon window and 2 mm PE cover with the 5 mm hole in diameter in the centre is used for PIGE analysis as well as for PIXE analysis of elements heavier than titanium. The energy resolution of 390 eV for 5.9 keV and 1.8 keV for 1.3 MeV (^{60}Co) makes it suitable for operation in energy regions of 4-75 keV and 50-1600 keV, respectively. The Si PIN photodiode Hamamatsu S3590 is used to register backscattered ions in front of the sample. A schematic view of the IBA chamber with detector positions is shown in Fig. 1.

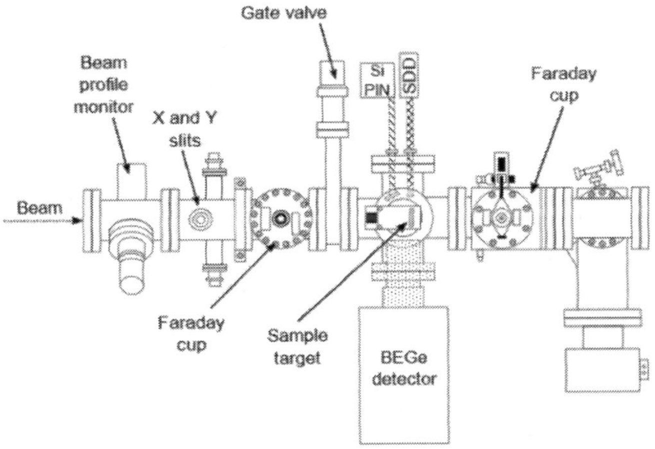

Fig. 1. Scheme of the IBA chamber in the CENTA laboratory.

3 MeV proton beams are frequently used for PIXE analysis with intensities ranging from 0.5 to 1.5 nA. For thin samples, the FC (Faraday cup) behind the sample is used with the current integrator Ortec 439 for charge determination. For thick samples, the charge can be estimated directly by its collection from the sample by insulated sample holder and current

integrator, or by periodic insertion of FC in front of the sample. Beam position on the sample is adjusted by the in-chamber camera and by recording a fluorescence on the scintillating surface. At least three spots on the surface of the sample were irradiated by protons during the PIXE measurements to estimate the homogeneity of the sample material.

2.2 Sample preparation

For the determination of the elemental concentration in the samples, the instrumental constant is needed. The so-called H-value was determined using thin MicroMatter standards in defined geometry. Due to different energy ranges, standard materials Fe, Cu, GaAs, CdSe, SrF_2, MoO_3, Sn, BaF_2, Au, Pb and MgF_2, GaP, KCl, were used for calibration of BEGe and SDD detectors, respectively.

Reference materials IAEA-158 Marine sediment and IAEA-359 Cabbage were used to validate the constructed H-values calibration. About 140 mg of reference material was mixed at the ratio of 80:20 with the Hoechst wax micro powder (Supelco) and pulverized by mortar and pestle. After thorough mixing, the samples were pelletized by KBr die with a diameter of 13mm and installed in the IBA chamber for evacuation.

3 Results and discussion

X-ray spectra from irradiated MicroMatter standards were processed by GUPIXWIN software, where the H-values were set as 1. Afterwards, the experimental H-value was calculated as the ratio between the calculated concentration from measured spectra and the nominal value. The energy dependence of H-values for both detectors is shown in Fig. 2. Because of complicated shapes, values were fitted by analytical functions in the range from 1.5 keV and 10 keV, for SDD and BEGe detectors, respectively. As can be seen, the H-value for the SDD detector is almost constant with a strongly decreasing trend below 1.5 keV, where the floating average was used instead of the fit. The same approach was used for values below 10 keV for the BEGe detector. This is the result of using C2 window transmissions for low Z elements (supplied by the manufacturer) during the calculation of thin standards concentrations. The trend of H-value energy dependency for the BEGe detector is different because we do not know the X-ray transmission factors through the detector window and the absorber for this detector. Therefore, together with geometry and efficiency, they were included in the final H-value with strong energy dependence.

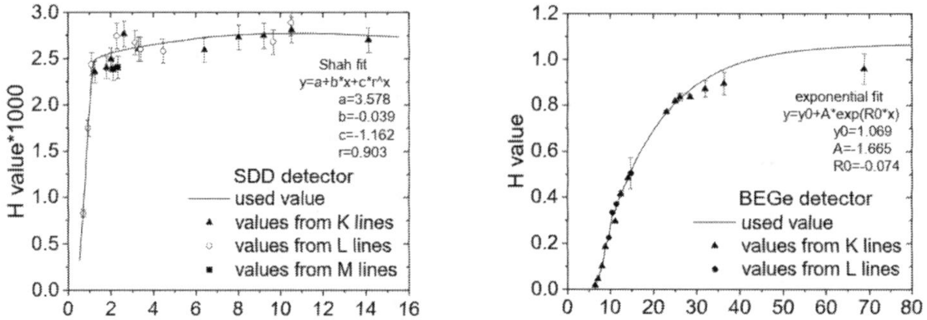

Fig. 2. Comparison of used and measured H-values for SDD (left) and BEGe (right) detectors.

To validate the H-value calibration of both detectors, the prepared thick samples of IAEA reference materials were measured in the same geometry. Elemental composition of samples was used for calculation of attenuation coefficient of each reference sample in GUPIXWIN with no additional correction. Calculated elemental concentrations of both reference materials are compared in Table 1 with their nominal values. The concentrations of elements below the detection limits of PIXE spectra were not compared.

As can be seen in Table 1, elemental concentrations of the IAEA-158 reference material are in good agreement with measured values, except for chromium, where a higher value was measured with SDD detector. The iron level in IAEA-359 was higher when compared with the reference value. The observed higher chromium and iron levels may be due to a contamination introduced by using a stainless-steel spoon during the sample removal from the mortar after its pulverizing. The iron contamination is not seen in the IAEA-158 sample due to its high iron content. Lower values of potassium and calcium observed in IAEA-359 may be due to inhomogeneities in the pressed samples.

Table 1. Elemental concentrations (mg kg^{-1}) with standard deviations for used reference materials

		Reference value	SDD	BEGe
IAEA-158 Marine sediment	Na	23800 ± 1000	26200 ± 2400	
	Al	51800 ± 3400	52400 ± 5200	
	K	20000 ± 1600	19400 ± 1400	
	Ca	64900 ± 5800	78600 ± 7900	
	V	73 ± 3.7	69 ± 49	
	Cr	74.4 ± 5.8	514 ± 36	
	Mn	356 ± 24	385 ± 35	
	Fe	26300 ± 1400	25800 ± 2000	25800 ± 2100
	Ni	30.3 ± 2.9	24 ± 7.6	
	Cu	48.3 ± 4.2	47 ± 23	
	Zn	140.6 ± 9.5	145.5 ± 19	174 ± 14
	Sr	473 ± 25	432 ± 86	502 ± 15
	Mg*	10390 ± 960	11040 ± 900	
	Ti*	3490 ± 170	3430 ± 270	
	Br*	224 ± 15	199 ± 50	214 ± 11
	Rb	82 ± 10		123 ± 11
	Ba	1028 ± 46		905 ± 64
IAEA-359 Cabbage	Na*	580 ± 21	790 ± 60	
	Mg	2160 ± 50	1960 ± 170	
	K	32500 ± 690	27800 ± 2700	
	Ca*	18500 ± 510	15900 ± 1500	
	Mn	31.9 ± 0.6	43 ± 17	
	Fe	148 ± 3.9	243 ± 10	338 ± 14
	Zn	38.6 ± 0.7	34.1 ± 10.2	41.5 ± 2
	Sr	49.2 ± 1.4		36.4 ± 2

*IAEA referred to as information values

4 Conclusion

For the successful use of the PIXE method in the CENTA laboratory, the energy dependence of the H-value was determined. The set of thin MicroMatter standards was used for calibration of SDD and BEGe detectors used for PIXE analysis. IAEA reference materials 158 (marine sediment) and 359 (cabbage) were used for validation of this calibration. Reasonable agreement was achieved between the reference and measured values for these reference materials. Higher concentrations of iron and chromium were determined in both samples, which probably resulted from contamination during the sample preparation. Therefore, the preparation method needs further improvements. Presented H-value dependence will be used for PIXE analysis of various samples analysed in the CENTA laboratory.

References

1. Z. Nejedlý, J.L. Campbell, W.J. Teesdale, C. Gielen, NIMB, **132** (1997)
2. M. Elfman, K.E. Limburg, P. Kristiansson, K. Malmqvist, J. Pallon, NIMB 150 (1999)
3. J. García López, F.J. Ager, M.B. Rank, F.J. Madrigaf, M.A. Ontalba, M.A. Respaldiza, M.D. Ynsa, NIMB **161** (2000)
4. I. Uzonyi, I. Rajta, L. Bartha, Á.Z. Kiss, A. Nagy, NIMB, **181** (2001)
5. J. Simičič, P. Pelicon, M. Budnar, B. Šmit, NIMB, **190** (2002)
6. O. Enguita, M.T. Fernández-Jiménez, G. García, A. Climent-Font, T. Calderón, G.W. Grime, NIMB **219-220** (2003)
7. M. Jakšić, I. Bogdanović Radović, M. Bogovac, V. Desnica, S. Fazinić, M. Karlušić, Z. Medunić, H. Muto, Ž. Pastuović, Z. Siketić, N. Skukan, T. Tadić, NIMB **260** (2007)
8. J. Meersschaut, J. Carbonel, M. Popovici, Q. Zhao, A. Vantomne, W. Vandervorst, NIMB, **331** (2014)
9. M.S. Rihawy, D. Halloum, Appl. Radiat. Isotopes, **133** (2018)
10. J.A. Maxwell, W.J. Teesdale, J.L. Campbell, NIMB, **95** (1995)
11. I. Harayama, D. Sekiba, Q. Zhao, A. Vantomme, W. Vandervorst, J. Meersschaut, NIMB, **406** (2017)
12. A. Shariff, K. Bülow, M. Elfman, P. Kristiansson, K. Malmqvist, J. Pallon, NIMB, **189** (2002)
13. T. Hassan, A. Shariff, A. Hossein, J. Abedin, A.K.M. Falzul Hoque, M.S. Chowdhuri, NIMA, **781** (2015)
14. P.P. Povinec, J. Masarik, P. Kúš, K. Holý, M. Ješkovský, R. Breier, J. Staníček, A. Šivo, M. Richtáriková, A. Kováčik, J. Szarka, P. Steier, A. Priller, NIMB, **342** (2015)
15. J. Zeman, M. Ješkovský, R. Kaiser, J. Kaizer, P.P. Povinec, J. Staníček, J. Radioanal. Nucl. Chem., **311** (2017)

The synthesis of Au-NPs by ion implantation in the crystalline GaN and characterisation of their optical properties

Adéla Jagerová[1,2*], *Josef* Flaks[1], *Zdeněk* Sofer[3], *Marek* Vronka[4], *Alena* Michalcová[5] and *Anna* Macková[1,2]

[1]Nuclear Physics Institute of the Czech Academy of Sciences, 250 68 Řež, Czech Republic

[2]Department of Physics, Faculty of Science, J.E. Purkyne University, Pasteurova 3632/15, 400 96 Ústí nad Labem, Czech Republic

[3]Department of Inorganic Chemistry, University of Chemistry and Technology Prague, Technicka 5, 166 28 Prague 6, Czech Republic

[4]Institute of Physics of the Czech Academy of Sciences, Na Slovance 1999/2, 18221 Prague, Czech Republic

[5]Department of Metals and Corrosion Engineering, University of Chemistry and Technology, Technická 5, 166 28 Prague, Czech Republic

Abstract. Nanostructured surfaces with embedded noble metal nanoparticles is an attractive way for manipulation with the optical properties of wide bandgap semiconductors applied in optoelectronics, photocatalytic processes or for Surface-Enhanced Raman spectroscopy. Ion implantation offers an effective way for nanoparticle preparation without the use of additional chemicals that offers precise control of nanoparticle depth distribution. The aim of this study is a synthesis of the gold nanoparticles in GaN by implantation of 1.85 MeV Au ions with high fluences up to 7×10^{16} cm^{-2} and study of optical properties of Au implanted GaN. Implanted crystals were annealed at 800 °C in an ammonia atmosphere for 20 min to support Au nanoparticle creation and GaN recovery. The structure characterisation has been realized by Rutherford backscattering spectroscopy in channelling mode and it showed the formation of two separated disordered regions – the surface region and buried layer. The lower implantation fluences induce damage mainly in a buried layer; however, the increase of the Au-ion fluence leads to the increase of surface disorder as well. Further, the increase of the Au-ion fluence induces the Au dopant shift to the surface and multimodal Au-depth profiles. TEM analyses confirmed the formation of Au nanoparticles in the implanted samples after annealing with sizes up to 14 nm. The increase of light absorption and modification of GaN bandgap of the Au modified GaN was deduced from the change in optical transmission spectra between 370 – 1400 nm.

*Corresponding author: jagerova@ujf.cas.cz

1 Introduction

Nowadays, noble metal nanoparticles (NPs) such as Au NPs, Ag NPs, or Pt NPs are a perspective way for manipulation with properties of crystalline semiconductors. The modification of GaN or ZnO wurtzite semiconductors with noble metal NPs can enhance their photocatalytic properties and modify their optical properties. The interaction of Au NPs with GaN crystals can result in the enhancement of the intensity of near-band edge photoluminescence emission [1]. Further, the presence of a plasmonic layer of Au NPs can significantly affect the efficiency of GaN as a substrate for Surface-Enhanced Raman spectroscopy (SERS) application, increase the intensity of SERS peaks and improve detection limits e.g. for cancer detection via miRNA solution [2,3].

The implantation of energetic metal ions with high fluence leads to the aggregation of dopant and can be applied as an interesting method for the preparation of metal-based NPs. Generally, this process involves several stages, depending on the implantation fluence, and it includes saturation of implanted ion concentration, further a start of dopant aggregation, nanoparticle growth, and formation of large particles at the expense of smaller ones known as an Ostwald ripening or coalescence of created nanoparticles into the compact layer. It is necessary to understand this process to maintain particle synthesis in the nanoscale dimension [4].

The ion beam method can be used for NPs preparation in selected depth, without the use of chemical additives [5]. However, it leads to high radiation damage, especially in the case of crystalline materials leading to the degradation of crystal properties. Therefore, there is still a need to study the implantation process with subsequent annealing for NPs preparation in connection to the optical response of prepared nanostructures.

The implantation of iron Fe^+ with energy 200 keV into the epitaxial layers of wurtzite gallium nitride (GaN) [6] leads to the formation of small clusters with a size 8 nm. The optical properties of lithium niobate ($LiNbO_3$) [7] were tailored by the Ag NPs prepared by implantation of Ag ions with energy 200 keV and fluences in the range of $5x10^{16}$ - $1x10^{17}$ cm^{-2}. Similar fluences were used also for the preparation of Cu NPs with diameter size 10 nm in single-crystalline silicon Si [8].

In this work, we implanted *c-plane* GaN single crystal substrates by energetic Au ions with a wide range of ion fluences to see particular stages of ion implantation induced NPs formation and their effect on the optical response of GaN crystal.

2 Experiment

The single crystal *c-plane* (0001) GaN substrates were implanted with 1.85 MeV Au ions with constant current density 270 nA/cm^2 at incidence 7° with high ion fluences in the range of 1.5×10^{16} cm^{-2} up to 7×10^{16} cm^{-2}, where the aggregation of Au ions and creation of Au NPs is expected. The projected range R_P of 1.85 MeV Au ions was estimated by the SRIM program as 229 nm with a standard deviation of $\Delta R_P = 64$ nm. The substrate holder was not directly cooled during the implantation experiment and substrate temperature increases up to 130 °C, as it was measured immediately after the implantation experiment. The 1.85 MeV Au ions were produced by the Tandetron accelerator in the Nuclear Physics Institute of CAS in Řež (Czech Republic). The implanted GaN substrates were annealed at 800 °C for 20 minutes in an ammonia atmosphere to support nucleation and growth of Au NPs. The structural characterisation of the implanted, as well as annealed GaN samples, were carried out by Rutherford backscattering spectroscopy (RBS-C) with a 2.8 MeV He^+ ion beam produced by Tandetron accelerator in Řež as well.

The visualization of prepared Au NPs was carried out on the implanted samples after annealling by transmission electron microscopy (TEM) measurement. Samples for transmission electron microscopy (TEM) were prepared as lamellas by focused ion beam SEM technique. HTEM (High-resolution transmission electron microscopy) and STEM (scanning transmission electron microscopy) observations were carried out by Fei Tecnai F20 field emission gun transmission electron microscope operated at 200 kV equipped with an EDS detector which confirmed the presence of Au NPs. For STEM, Z-contrast imaging was performed using a high angle annular dark field (STEM–HAADF) detector.

The optical response of the implanted GaN substrates before and after annealing was analysed by measurement of transmission spectra in the UV-Vis region. Transmission spectra were measured perpendicularly to the GaN surface with UVISEL plus Spectroscopic Ellipsometer from Horiba (France) in the spectral range 190-1400 nm with a step 5nm.

3 Results and discussion

As it is seen in RBS-C spectra presented in
Fig. 1a, the implantation of Au ions leads to the increase of yield of backscattered ions from Ga (channels up to 540) even for the lowest Au-ion fluence 1.5×10^{16} cm^{-2}. The backscattered ion yield in RBS-C spectra is connected to the growth of Ga-sublattice disorder and the displacement of Ga atoms from their original positions. The ion implantation of GaN crystals forms two damage regions. Firstly, the buried layer corresponds to the depth of implanted ions (channels 410 – 460). An increase in implantation fluence leads to the saturation of damage in the buried layer. The damage level characterised by the normalized yield χ_{norm} achieved more than 71 % for the sample implanted with fluence 1.5×10^{16} cm^{-2}, for the higher fluences saturates at around 80 %. Defect migration towards the surface is forming a second surface damage region (channels 460 – 525). In this region, χ_{norm} gradually increases from 56 % for the lowest implantation fluence up to 95 % for the fluence 7×10^{16} cm^{-2}. For the highest implantation fluence 7×10^{16} cm^{-2}, the yield of backscattered ions in the surface damage region reaches a level of random spectra indicating nearly complete amorphisation of the surface or creation of randomly oriented nanocrystals and voids [9,10].

The low temperature (200 K) 2 MeV Au GaN implantation, with the high implantation fluences leads to the significant change in crystal damage and to the creation of surface and buried amorphous layers for fluences above 1.4×10^{15} cm^{-2} as is presented in [11]. In our experiment, only the surface amorphous region was observed. During the ion implantation at room temperature, GaN exhibits strong dynamic annealing which prevents bulk crystals to reach complete amorphisation in a depth. Instead of this, GaN amorphisation proceeds layer-by-layer from the GaN surface [12]. Further, the high ion fluence bombardment tends to induce preferential nitrogen loss from the GaN surface [12] as it was also observed in this experiment by an increase of Ga amount in the surface layer in a random spectrum of the sample implanted with the highest fluence 7×10^{16} cm^{-2} (see **Fig. 1a**). The post implantation annealing in the ammonia atmosphere leads to crystal recovery and reorganization mainly in the surface region, as is seen in **Fig. 1b**. The annealing in an atmosphere with an excess of nitrogen can protect the GaN surface against disintegration and compensate for nitrogen loss by the creation of new Ga-N bonds during high-temperature annealing [13].

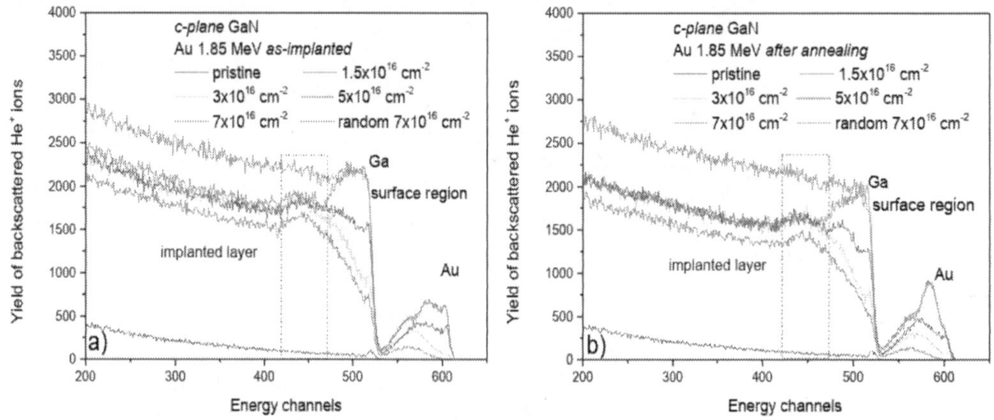

Fig. 1. RBS-C spectra of GaN pristine as well as the samples implanted with Au ions with energy 1.85 MeV in a), and corresponding samples after 20 minutes of annealing at 800 °C in ammonia atmosphere in b)

RBS-C analysis serves also for the Au-ion depth profiling which was evaluated from random spectra. In **Fig. 2a** we can see a comparison of SRIM calculated Au-depth distribution to the experimentally determined Au-depth profiles after the ion implantation. For the lower implantation fluences, up to 3×10^{16} cm^{-2}, the Au depth distribution has a Gaussian profile and agrees with SRIM prediction. With an increase of ion implantation fluence, the Au starts to migrate towards the GaN surface and create multimodal depth profiles with additional concentration maxima. It was observed that Au ions diffuse to the GaN surface region where the high disorder nearly amorphous structure is present [11,14]. Annealing induces reorganization of Au depth profiles for the samples implanted with fluence 5×10^{16} cm^{-2} and higher, as is seen in **Fig. 2b**. The most significant change of Au depth distribution is observed for the sample implanted with the highest Au-ion fluence, where a sharp Au-concentration maximum appeared in the depth of 125 nm. The integrated Au-content stays constant before and after annealing.

It was shown that thermal or laser annealing can significantly affect the formation process of metal NPs. The annealing support movement of implanted Au ions towards the region with extended defects serving as trapping sites for implanted ions [15]. This leads to the dopant depth redistribution and enhancement of Ostwald ripening and coalescence of NPs [15]. Such phenomenon is suggested to take place also in our case, where probably Au-redistribution and NPs coalescence take a place.

https://doi.org/10.1051/epjconf/202226101003

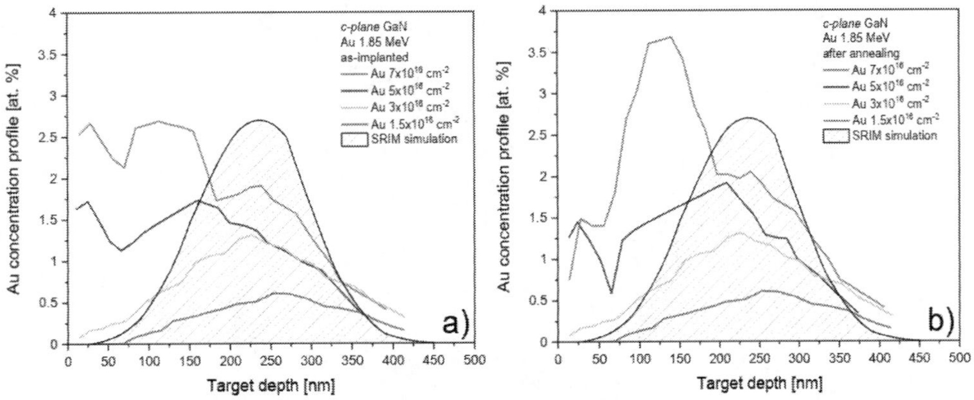

Fig. 2. – The comparison of SRIM calculated Au-depth distribution corresponding to the fluence 5×10^{16} cm^{-2} of 1.85 MeV Au ions with the experimentally determined Au-concentration depth profiles in a) and after the subsequent annealing in b)

TEM analysis was performed on the implanted GaN samples after the thermal annealing. Implanted Au ions tend to aggregate and create small clusters with sizes up to several nm up to the depth of almost 300 nm, see cross-sectional STEM image in **Fig. 3a**. These small Au NPs have a diameter in a range of 4 - 15 nm. The STEM image of GaN implanted with ion fluence 7×10^{16} cm^{-2} presented in **Fig. 3a** is superimposed with Au ion depth distribution obtained from RBS-C. The biggest Au NPs are distributed in depth from 70 nm up to 180 nm corresponding to the maximum of Au distribution determined by RBS-C. Above this depth, the amount of Au NPs decreases and large Au agglomerates are presented up to 300 nm. The dark spots belong to extended defects and voids in the highly damage region of GaN implanted with 2 MeV Au ions observed in [11]. The structure and lattice spacing in Au NPs were determined by HRTEM analysis with Fast Fourier Transformation (FFT); see **Fig. 3b** and **Figure 3c**. Prepared Au NPs exhibits a face-centred cubic (*fcc*) structure with d-spacing of 0.38 nm. The *fcc* structure is typical for pure gold and it was already observed for Au NPs prepared by the ion implantation of Au ions to the silica glass [16] as well as for Au NPs colloid solution prepared by standard chemical preparation procedure from Au salts [17]. In our experiment, the d-spacing of prepared Au NPs is lower than the d-spacing of pure Au 0.41 nm [17,18] which may be attributed to internal strain.

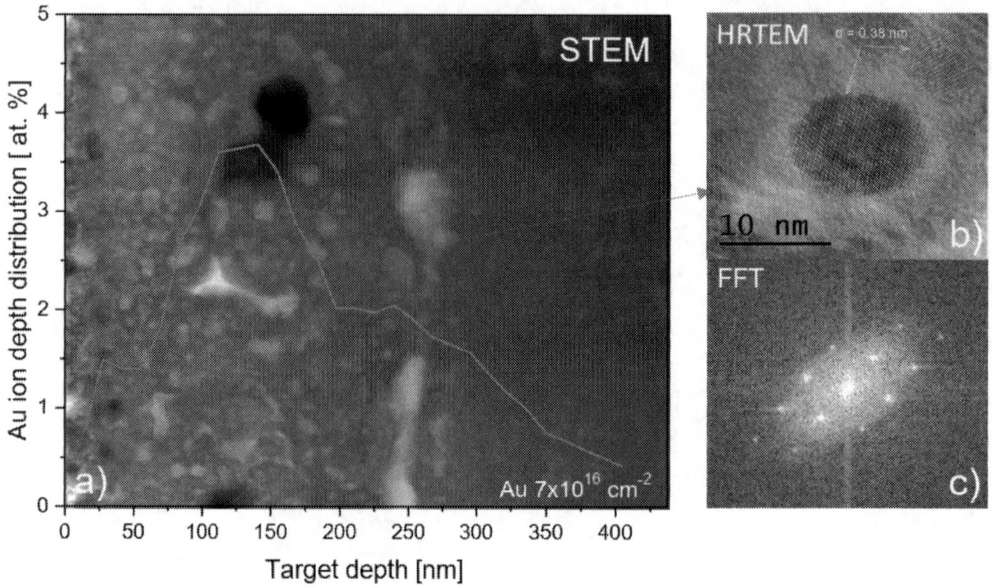

Fig. 3. Cross-sectional STEM image of GaN sample implanted with the Au-ion fluence 7×10^{16} cm^{-2} after annealing in a) and HRTEM analysis of Au NPs with diffractogram in b) and corresponding FFT pattern of the bigger Au NPs.

The optical response of the Au-ion implanted, as well as subsequently annealed, GaN samples was characterized by optical transmission spectra in a range 190-1400 nm presented in **Fig.4**. The GaN is transparent for photons with energy lower than GaN bandgap energy (3.4 eV [19]). Therefore, the pristine GaN exhibits transmission above 60 % for light with a wavelength higher than 370 nm. The light with a wavelength lower than 370 nm is absorbed in GaN. Optical transmission in the wavelength range 370 – 1400 nm is influenced by the Au-ion implantation, generally, the optical transmission decreases with the growing Au-ion implantation fluence. This fact is connected to the higher crystal disorder. For the highest Au-ion fluence 7×10^{16} cm^{-2}, the optical transmission is strongly suppressed in the wavelength above 370 nm and the sample appears nearly opaque. Additionally, the transmission edge is shifted towards higher wavelengths. This indicates the lowering of bandgap energy with implantation fluence, as is presented in **Fig. 4a**. The modification and lowering of bandgap energy can be attributed to the strain caused by incorporated Au ions and growing structure disorder as it was already observed for a similar structure of ZnO modified with Au NPs [20].

The post implantation annealing leads to the crystal recovery as it was observed by RBS-C and it results in an increase of the optical transmission in the wavelength range of 370 - 1400, see **Fig. 4b.** The recovery effect is seen especially for the lower implantation fluences up to 3×10^{16} cm^{-2}, where the transmittance spectrum reaches the level of the pristine GaN above the wavelength of 800 nm. The shift of the transmission edge towards higher wavelengths indicates a decrease of bandgap energy due to crystal recovery and the lowering of lattice strain after the annealing. Nevertheless, the position of the transmission edge still indicates some persisting damage even after the annealing. Further, the Au NPs are characterised by increased absorption due to the Surface plasmon resonance (SPR) leading to the creation of a dip between 500 – 600 nm in transmission spectra [20], which is not observed in our experiment. It can be assumed that Au NPs are located in a depth of GaN and their volume fraction in GaN samples is too low to allow propagation of SPR response in the transmission spectra similarly as it was evidenced for Cu NPs synthesised by ion implantation in ITO layers in [21].

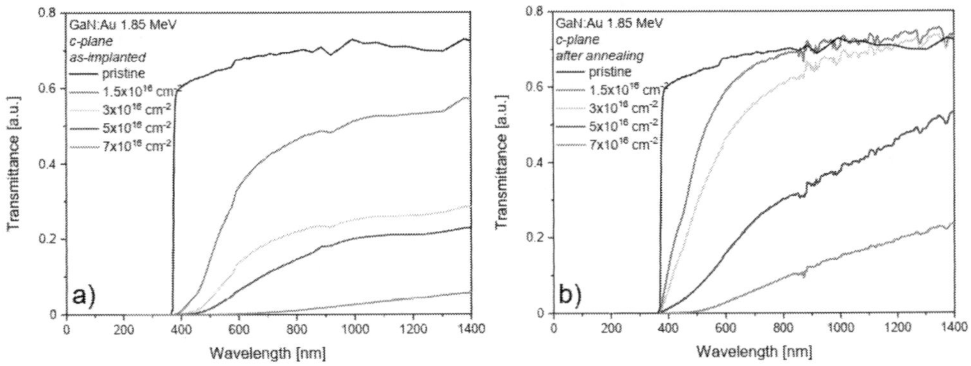

Fig. 4. Optical transmittance spectra of the pristine and Au-ion implanted GaN in a) and the subsequently annealed GaN samples in b)

4 Conclusion

The ion implantation of 1.85 MeV Au ions into the *c-plane* GaN crystals followed by thermal annealing at 800°C in an amonia atmosphere has been applied for the preparation of Au NPs to modify GaN optical properties. The Au-ion implantation with high fluences leads to the Ga-sublattice disorder enhancement in the implanted layer as well as to the additional disordered layer created at the surface with thickness about 150 nm for the Au-ion fluence above 5×10^{16} cm^{-2}. Above this Au-ion fluence, the creation of the Au-multimodal depth profiles was observed. The annealing induced partial crystal recovery in the implanted layer and a significant improvement of the surface layer ordering as it was deduced from RBS-C analysis. Simultaneously, the implanted Au depth profiles exhibited a sharp Au-concentration maximum which may be related to the Oswald ripening and growth of Au NPs. The TEM analysis confirmed the formation of small Au NPs of the sizes 1 – 15 nm in the implanted GaN after annealing. TEM showed Au NPs exhibiting cubic *fcc* structure with d-spacing of 0.38 nm. Au-ion implantation leads to the suppression of the optical transmission above 370 nm with a growing Au-ion implantation fluence. This effect is partially recovered after the annealing and mainly for the Au-ion fluences below 5×10^{16} cm^{-2}.

The research has been carried out at the CANAM (Centre of Accelerators and Nuclear Analytical Methods) infrastructure LM 2015056. This publication has been supported by OP RDE, MEYS, the Czech Republic, under the project CANAM OP, CZ.02.1.01/0.0/0.0/16_013/0001812 and promoted by the University of J.E. Purkyne project UJEP-SGS-2021-53-002-2

References

1. I.-H. Lee, L.-W. Jang and A. Y. Polyakov, *Nano Energy*, **13**, 140–173 (2015)
2. Y. Han, L. Qiang, Y. Gao, J. Gao, Q. He, H. Liu, L. Han and Y. Zhang, *Applied Surface Science*, **541**, 148456 (2021)
3. I. Dzięcielewski, J. Smalc-Koziorowska, M. Bańkowska, T. Sochacki, A. Khachapuridze and J. Weyher, *Applied Surface Science*, **378**, 30–36 (2016)
4. F. Chen, H. Amekura and Y. Jia, in *Ion Irradiation of Dielectrics for Photonic Applications*, Springer Singapore, Singapore, vol. 231, pp. 61–107 (2020)

5. M. E. Koleva, N. N. Nedyalkov, Ru. Nikov, Ro. Nikov, G. Atanasova, D. Karashanova, V. I. Nuzhdin, V. F. Valeev, A. M. Rogov and A. L. Stepanov, *Applied Surface Science*, **508**, 145227 (2020)
6. G. Talut, H. Reuther, A. Mücklich, F. Eichhorn and K. Potzger, *Appl. Phys. Lett.*, **89**, 161909 (2006)
7. C. Pang, R. Li, Y. Zhang, Z. Li, N. Dong, L. Wu, H. Yu, J. Wang, F. Ren and F. Chen, *Opt. Express*, **26**, 31276 (2018)
8. A. I. Gumarov, A. M. Rogov and A. L. Stepanov, *Composites Communications*, **21**, 100415 (2020)
9. F. Gloux, T. Wojtowicz, P. Ruterana, K. Lorenz and E. Alves, *Journal of Applied Physics*, **100**, 073520 (2006)
10. M. Ishimaru, Y. Zhang and W. J. Weber, *Journal of Applied Physics*, **106**, 053513 (2009)
11. Y. Zhang, M. Ishimaru, J. Jagielski, W. Zhang, Z. Zhu, L. V. Saraf, W. Jiang, L. Thome and W. J. Weber, *J. Phys. D: Appl. Phys.*, **43**, 085303 (2010)
12. S. O. Kucheyev, J. S. Williams and S. J. Pearton, *Materials Science and Engineering: R: Reports*, **33**, 51–108 (2001)
13. C. Ronning, *Physics Reports*, **351**, 349–385 (2001)
14. A. Macková, P. Malinsky´, A. Jágerová, Z. Sofer, D. Sedmidubský, K. Klímová, R. Böttger and S. Akhmadaliev, *Surf Interface Anal*, **50**, 1099–1105 (2018)
15. M. Steinert, W. Wesch, A. Undisz, M. Rettenmayr, W. C. Nunes, R. P. Borges, M. Godinho, R. M. Rubinger, M. C. Carmo and N. A. Sobolev, *J. Phys. D: Appl. Phys.*, **42**, 035406 (2009)
16. H. G. Silva-Pereyra, J. Arenas-Alatorre, L. Rodriguez-Fernández, A. Crespo-Sosa, J. C. Cheang-Wong, J. A. Reyes-Esqueda and A. Oliver, *J Nanopart Res*, **12**, 1787–1795 (2010)
17. D. Lomelí-Marroquín, D. Medina Cruz, A. Nieto-Argüello, A. Vernet Crua, J. Chen, A. Torres-Castro, T. J. Webster and J. L. Cholula-Díaz, *IJN*, **14**, 2171–2190 (2019)
18. O. Peña, U. Pal, L. Rodríguez-Fernández, H. G. Silva-Pereyra, V. Rodríguez-Iglesias, J. C. Cheang-Wong, J. Arenas-Alatorre and A. Oliver, *J. Phys. Chem. C*, **113**, 2296–2300 (2009)
19. Y.-T. Chen, C.-Y. Yang, P.-C. Chen, J.-K. Sheu and K.-H. Lin, *Sci Rep*, **7**, 5788 (2017)
20. R. Sreeja Sreedharan, V. S. Kavitha, S. Suresh, R. Reshmi Krishnan, R. Jolly Bose and V. P. Mahadevan Pillai, *Appl. Phys. A*, **124**, 815 (2018)
21. A. I. Ryasnyansky, B. Palpant, S. Debrus, R. I. Khaibullin and A. L. Stepanov, *J. Opt. Soc. Am. B*, **23**, 1348 (2006)

The structural and optical response of the Au nanoparticles embedded in YSZ modified using high-energetic ion irradiation

Romana Mikšová[1*], *Jakub* Cajzl[3], and *Anna* Macková[1,2]

[1]Nuclear Physics Institute of the Czech Academy of Sciences, p.r.i, 250 68 Řež, Czech Republic

[2]Department of Physics, Faculty of Science, J.E. Purkyně University, Pasteurova 3544/1, 400 96 Ústí nad Labem, Czech Republic

[3]Department of Inorganic Chemistry, University of Chemistry and Technology in Prague, 166 28 Prague, Czech Republic

Abstract. Gold nanoparticles (Au NPs) were synthesized in single-crystalline yttria-stabilized zirconia (YSZ) using a high-energetic gold ion implantation with the intention of subsequent modification by silicon ion irradiation. (100)-oriented YSZ samples were implanted with 1 MeV Au^+ ions at room temperature and fluences of 1.5×10^{16} cm^{-2}, 5.0×10^{16} cm^{-2} and 7.5×10^{16} cm^{-2} and subsequently annealed for 1 h at 1100 °C in air. Gold NPs modification was realized with irradiation by 10 MeV Si^{3+} with the fluence of 5.0×10^{14} cm^{-2}. Au NPs distribution and subsequent YSZ structure modification were studied. YSZ samples as-implanted with Au ions, annealed and subsequently irradiated with Si ions were characterized using Rutherford backscattering spectrometry (RBS) and RBS in channelling mode (RBS-C). The RBS spectra show that the implanted Au-ions concentrate in the YSZ sub-surface layer in the depth of about 50-250 nm and shift to greater depth with increasing ion fluence. The Si^{3+} ions irradiation slightly increased Zr sub-lattice disorder. Optical absorbance shows an increase of absorption band at 550 nm after Au implantation only for the two highest ion fluences. After annealing and Si irradiation, optical absorbance increased at 530-580 nm with connection to ion implantation fluence of Au^+ ions.

1 Introduction

Gold nanoparticles (Au NPs) embedded in oxide matrices exhibit unique optical properties that can be used for a lot of applications, such as high-density information storage devices, ultrafast optical switches, and optical gas sensors [1]. The reliable production of such optical devices requires the ability to customize the optical response of the nanocomposite systems. In the case of gold (Au), the optical response of Au NPs is dominated by a surface plasmon resonance band (SPR), which exists due to the light-induced displacement of

* Corresponding author: miksova@ujf.cas.cz

conduction electrons to the positive ionic background in the NP [2]. The optical response of Au NPs is influenced by their shape, size, geometrical configuration, and composition [3]. The ion irradiation of the existing nanocomposite system can be used for the deformation of the NPs [4]. The threshold for Au NPs modification by subsequent energy transfer from energetic ions (electronic stopping power S_e), below which the Au nanoclusters remained spherical, depends on the particle size as well as on the substrate structure. For instance, modifications of Au NPs in silica with d =15 nm occur during irradiations by energetic ions with S_e of 3.5 keV/nm [5] and higher. For Au NPs in silica with 40–50 nm diameter, the electronic stopping power above 5.5 keV/nm is required for elongation. The elongation of Au NPs in silica with 80 nm diameter is observed for S_e ~7–8 keV/nm [6] (irradiated with high-energetic Ag and Cu ions). A similar value of 3.3 keV/nm was found in reference [5] for silica shell-Au core colloidal particles with identical sizes. In reference [7], a value of (2.0±0.5) keV/nm was deduced for polydisperse Au particles in silica with d < 14 nm.

Ion implantation has an important role in the fabrication of nanoclusters embedded in materials. After heating to temperatures high enough that the implanted atoms become mobile, they will segregate from the matrix and form nanoclusters if the implantation depth is suitable compared to the migration distance [8]. Yttria-stabilized zirconia (YSZ) is widely used for advanced technological applications, such as thermal coating [9], gas sensors [10], or colourless ceramics in the jewellery [11] due to its distinctive features like high chemical stability and good radiation resistance. YSZ is also used as oxygen ion conductors, with applications in solid oxide fuel cells, oxygen pumps, and oxygen sensors. In these applications, YSZ functions as a solid electrolyte, where oxygen ions diffuse freely if the operating temperature is higher than required for the diffusion of oxygen ions (T > 300 °C) [12]. By adding Au NPs to YSZ, the diffusion kinetics of oxygen ions into the oxygen vacancies in the YSZ lattice is expected to increase as in general metals increase electrochemical reaction kinetics at the surfaces of doped YSZ electrodes [13].

This study focuses on the preparation of Au NPs created in YSZ by ion implantation using 1 MeV Au-ions and the subsequent irradiation is performed by 10 MeV Si^{3+}. It was shown that noble metals nanoparticles (Ag) can be effectively synthesized in YSZ by ion implantation [14]. Subsequent annealing was used for nanoparticle formation and the creation of Au: YSZ nanocomposite layers. Defect recovery in the nanocomposite layer will be discussed in connection to the annealing procedure and the additional energetic Si^{3+} ion irradiation. The structural and optical response of the prepared nanocomposite system in YSZ has been investigated using Rutherford backscattering spectrometry (RBS), RBS in channelling configuration (RBS-C), and optical measurement. While this methodology cannot replace the high-resolution analysis of implanted samples, it does offer a non-destructive, suitable, and cheaper alternative and complementary route for microscopy analysis when available.

2 Experimental set-up

The (100)-oriented YSZ fully stabilized with 9.5 mol. % Y_2O_3 (containing 1.5 wt. % Hf as the contaminant) were implanted with 1 MeV Au^+ ions at room temperature and fluences of 1.5×10^{16} cm^{-2}, 5.0×10^{16} cm^{-2} and 7.5×10^{16} cm^{-2}. An off-axis 7° regime to minimize the channelling effect and non-controllable Au dispersion into a deeper layer was used. An ion current density lower than 0.2 μA.cm^{-2} was used to avoid target heating during implantation. Subsequent annealing in the air was performed for 1 h at 1100 °C. The irradiation by 10 MeV Si^{3+} (Si^+ listed below for simplicity) with the fluence of 5×10^{14} cm^{-2} was attempted to modify Au NPs shapes and distribution. All implantation experiments were carried out in the implantation chamber at the Tandetron Laboratory of the Nuclear

Physics Institute (NPI) in Řež (Czech Republic). A working pressure below 5.3×10^{-5} Pa was maintained by the pumping system based on a turbomolecular pump.

The Au concentration–depth profiles and structural modification were studied by Rutherford backscattering spectrometry (RBS) and RBS in channelling mode (RBS-C) using the beam of 2.8 MeV He^+ ions at the laboratory scattering angle of 170° with the depth probed of RBS ~2 μm (in this geometry). Higher energy was used to differentiate Zr and Au signals in the RBS spectra. Random and aligned spectra (the spectrum of back-scattered ions with the ion beam incident along the crystallographic axis) were recorded using the defined rotation of the single-crystal sample which was mounted on the two-axis goniometer allowing for the rotation with a resolution of 0.01°. The ion currents used during the RBS analysis were ~5 nA. All samples were measured in three different spots to avoid additional damage during the measurements. The elemental concentration was analysed using the SIMNRA code [15].

The SRIM code [16] was used for the theoretical predictions of ion interaction with matter. Monte Carlo (MC) simulations performed with SRIM provided the calculated Au depth distribution in YSZ. The calculated results were compared to the experimental data. For a better understanding of electronic and nuclear stopping interplay in the implanted and subsequently irradiated YSZ, the SRIM simulations of energy losses were performed. Electronic and nuclear-energy stopping at various depths was calculated by simulating 1 MeV Au and 10 MeV Si penetrations through YSZ. The ratio between electronic and nuclear energy losses integrated into the whole penetration depth of ions was calculated as $S_e/S_n = 0.3$ (Au) and $S_e/S_n = 237$ (Si). In the surface region, the electronic stopping of Si ions predominate over nuclear ($S_e \sim 6.3$ keV/nm). Projected ranges and standard deviations of 1 MeV Au ions and 10 MeV Si were simulated by SRIM as follows: $R_P = 134$ nm and $\Delta R_P = 40$ nm, $R_P = 2.65$ μm and $\Delta R_P = 225$ nm, respectively.

Optical absorption spectra were measured using Avantes AvaSpec-ULS-3648 Starline high-resolution fibre-optic spectrometer with linear CCD detector in the range of 250–600 nm at room temperature. As a source, a combined deuterium-halogen lamp Avalight-DHc was used. The light was guided from the source lamp using multimode optical fibre to the holder with the sample and the transmitted light was then collected with another multimode low-dispersion optical fibre into the detector. Optical absorption spectra were smoothed for better clarity and normalized to baseline in OriginPro 9.

3 Results and discussion

3.1 RBS-C method

The comparison of the aligned and the random RBS-C spectra of the pristine YSZ with as-implanted YSZ at the ion fluences of 1.5×10^{16} cm^{-2}, 5.0×10^{16} cm^{-2} and 7.5×10^{16} cm^{-2} are shown in Figure 1a–c for the (100)-oriented YSZ sample, respectively. The random spectra exhibited in the high backscattering energy parts decrease in Zr signals. This decrease is connected to the presence of Au atoms confined in the thin region near the surface. This phenomenon is more pronounced after the Au-ion implantation at the highest fluence. In Figure 1, the decrease of Au signal in the aligned spectrum is connected to any Au preferential position in YSZ or shadowed with Zr atoms. A similar trend was observed in our previous work in ZnO [17]. Two disordered regions in the aligned RBS spectrum are shown: (i) the surface region connected to the modified YSZ surface region (~544–558 channels), and (ii) the deeper damage region (~490–530 channels) created after Au implantation (highlighted in the black rectangle). Figure 1 shows that the damaged region is in a higher depth than is projected range of Au simulated by SRIM. It can be connected to defect migration during the ion implantation.

(i) In the surface damage region, a significant decrease of the Zr-disorder after the annealing for the two lower fluences was shown (Figures 1a and 1b – green line). After Si-ion irradiation disorder increases again (Figures 1a and 1b – blue line). In the case of 7.5×10^{16} cm^{-2}, annealing did not contribute to the reconstruction of the surface damage region; on the contrary, an increase in damage was shown (see Figure 1c – green line). In subsequent Si^{+} irradiated samples, the decrease of the disorder on the surface region was observed (see Figure 1c – blue line).

(ii) The damage accumulation in the deeper damage region was monitored using the normalised yields (χ_z) summarised in Figure 1d and Table 1 as a function of the Au-ion fluence. The values of χ_z are extracted from the RBS-C spectrum as the ratio of the yield in the aligned spectrum to the yield in the random spectrum in the highlighted region of interest (black rectangle) connected directly to Zr-sub-lattice disordering. The average value of χ_z for each sample was calculated from three measurements of χ_z as well as the statistical error of the mean $_{\Delta}\chi_z$ [18].

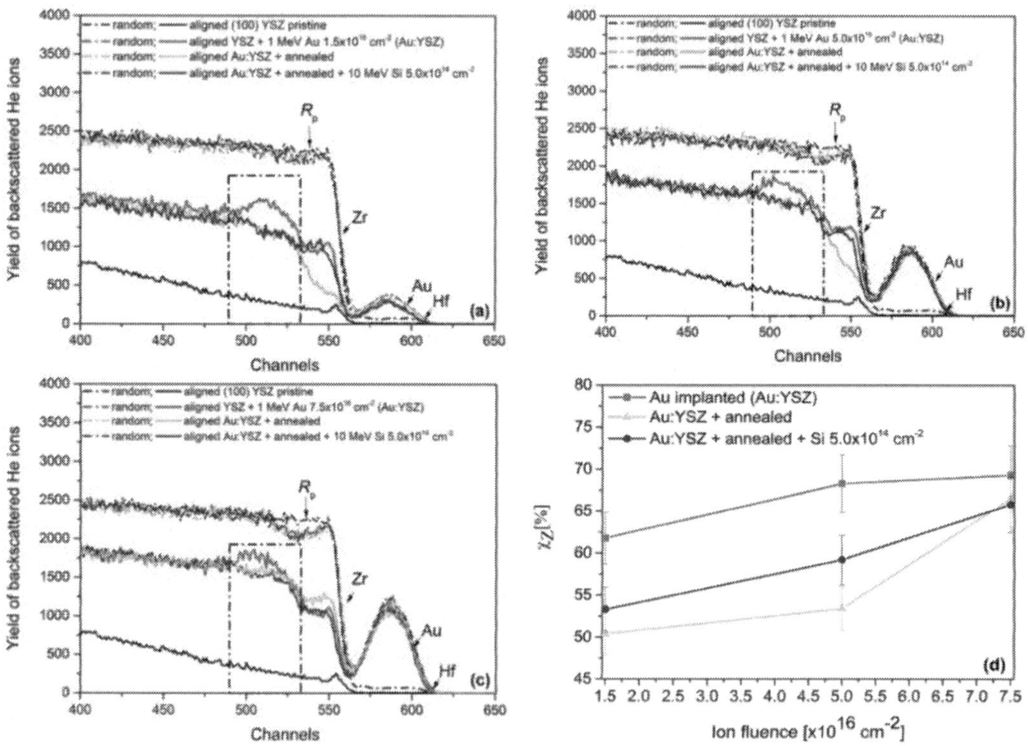

Fig. 1. The RBS-C spectra of the pristine, Au-implanted (with ion fluences of 1.5×10^{16} cm^{-2} (a), 5.0×10^{16} cm^{-2} (b) and 7.5×10^{16} cm^{-2} (c)), annealed and subsequently Si-irradiated YSZ at ion fluence of 5.0×10^{14} cm^{-2} for (100) orientation. The normalised yields (χ_z) are summarised as the function of Au^{+}-ion fluence (d).

The increase of χ_z in the deeper damage region was observed with the increasing Au^{+} ion fluence (Figure 1d – red line). After annealing, the crystalline structure of YSZ is partially healed (Figure 1d – green line); the Zr-disorder decrease is more pronounced for the lower

Au-ion implantation fluences, as was expected. Si-ion irradiation almost did not influence the value of χ_z in the deeper damage region (see Table 1).

The Au concentration-depth profiles were obtained in the Au-implanted, annealed and Si-irradiated samples from the RBS random spectra. The experimental Au concentration-depth profiles exhibit the normal distribution shape for all ion fluences, all projected ranges R_P and the standard deviation are shown in Table 1.

Tab. 1. The mean projected range R_P and the standard deviation ΔR_P deduce from the experimentally Au concentration-depth profiles from RBS. χ_z for YSZ implanted with Au-ions, annealed and subsequently irradiated with Si ions are presented in the last column.

(100) YSZ	Au implanted		annealed		Si irradiated	
Fluence [Au ions.cm^{-2}]	$R_P \pm \Delta R_P$ [nm]	$\chi_z \pm \Delta\chi_z$ [%]	$R_P \pm \Delta R_P$ [nm]	$\chi_z \pm \Delta\chi_z$ [%]	$R_P \pm \Delta R_P$ [nm]	$\chi_z \pm \Delta\chi_z$ [%]
1.5×10^{16}	151±45	61.8±3.1	156±46	50.4±2.5	157±46	53.3±2.7
5.0×10^{16}	161±63	68.3±3.5	169±62	53.4±2.7	168±62	59.2±3.0
7.5×10^{16}	184±73	69.3±3.6	181±72	66.5±3.4	180±70	65.8±3.2

The Au-concentration depth profiles are different from the calculated values with the projected range R_P calculated by SRIM-2013 as $R_P \pm \Delta R_P$ = 134±40 nm. The experimental values of R_p and ΔR_p increase with the Au-ion implantation fluence. It was observed additional shifts of Au-maximum concentration to greater depth after the annealing, but after Si-ion irradiation, R_p is almost the same as after annealing (see Table 1). The maximum is shifted by about 30 % to the depth (~50 nm) for the Au implanted samples with the highest ion fluence. A similar trend was found in our previous work in Si implanted by Au^+ ions [19], where the maximum shift is about 23 % (~35 nm). Discrepancies in the experimental Au-concentration depth profiles can be connected to overestimating electronic stopping in Au implemented in SRIM. It was mentioned in the study [20] that for heavy ions at low energy, the measurements of stopping show that SRIM electronic stopping is higher than the experimentally measured values. Nevertheless, there must be also taken into account the gradual modification of the YSZ structure during the implantation. Such a large shift, especially for the highest fluence, points to a certain Au migration.

3.2 Optical characterisation

Optical absorption spectra of the pristine YSZ samples (see Figure 2) showed the main absorption band around 280 nm (i.e. approx. 4.4 eV). From the literature, the optical band gap in YSZ, i.e. the energy transition between conduction and valence band, is 4.4–5.4 eV, i.e. 230–280 nm [21]. Optical absorption spectra of the samples Au-implanted YSZ are depicted in Figure 2a as well as results for samples, which were subsequently annealed at 1100 °C and Si-irradiated. The zoom of the region of interest in 450–680 nm wavelengths, where SPR peaks are localized, is presented in Figure 2b. The increase of the absorption band at around 550 nm was observed in Au implanted samples with fluence higher than 1.5×10^{16} cm^{-2}.

https://doi.org/10.1051/epjconf/202226101004

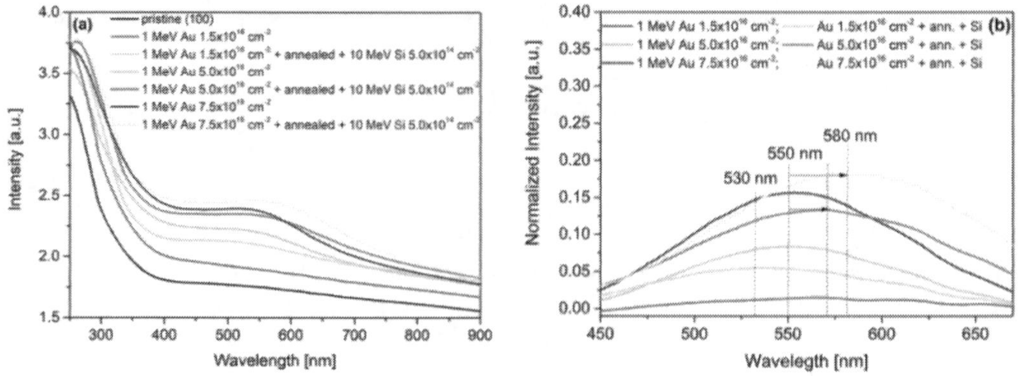

Fig. 2. Optical absorption spectra of the Au-implanted (100)-oriented YSZ (Au: YSZ) and Au: YSZ, annealed and Si-irradiated in the wavelength range from 250 to 900 nm and the zoom of the region of interest in 450-680 nm wavelengths is presented (b).

After annealing and Si-ion irradiation, the situation changed and the increase of the absorption band at the wavelength ~530 nm was observed for the lowest Au-ion fluence of 1.5×10^{16} cm^{-2} and at the wavelength between 570–580 nm for the two higher Au-ion fluences. This absorbance band can be attributed to gold nanoparticles (NPs) formation in YSZ [22,23], where the central wavelength can vary depending on the size of Au NPs. With increasing Au-ion implantation fluence, the absorption band centred at 550 nm increased as well. It can be connected to the enhanced nanoparticle creation (SPR activity) in the YSZ implanted and annealed layer, since the nanoparticle number increases with the increasing Au-ion fluence. However, we have to mention that the absorbance in the whole wavelength range increased as well, the YSZ samples became less transparent to visible light, which can be attributed to the creation of various extrinsic and intrinsic defects (vacancies, interstitials, etc.) in the structure of YSZ samples (defects caused the creation of new energy levels inside bandgap for optical absorption transitions). After the Si-ion irradiation, the absorbance band is shifted to the longer wavelengths about 580 nm and intensity increased. It can be ascribed to enhancing crystalline disorder as well as Au NPs clustering, or Au NPs size growth.

3.3 Discussion

(100)-oriented YSZ crystals were implanted with 1 MeV Au^+ ions at ion fluences of 1.5×10^{16} cm^{-2}, 5.0×10^{16} cm^{-2} and 7.5×10^{16} cm^{-2} and subsequently annealed at 1100 °C in the air for 1 hour. The Si-ion irradiation with energy 10 MeV was used to modify the created Au NPs. RBS, RBS-C and optical measurements on the produced samples were performed. It is shown from RBS, that the Au becomes distributed over a larger depth range with increasing Au ion fluence which can be connected to Au ion migration during ion implantation. The RBS-C showed higher damage in Zr-sublattice up to the projected range of Au ions in YSZ after the ion implantation. Subsequent annealing caused only partial healing of the crystalline YSZ structure, especially in the surface region (Figure 1a and 1b) and two lower ion fluences. 10 MeV Si-ion irradiation increased the disorder in the surface damage region, no additional disorder was observed in the deeper damage region. In the case of the highest Au ion fluence, the annealing caused the increase of disorder in the surface region and conversely reduced the damage in the surface region after Si-ion irradiation (Figure 1c).

The optical measurement showed the presence of Au NPs. Their growth was more pronounced for the two higher Au-ion fluences, where SPR peaks at 550 nm were shown.

No SPR peak was found for the lowest ion fluence. The band was previously connected with two defect types: oxygen vacancies with trapped electrons (F-type centres) nearest to Zr^{4+} cations and hole trapped oxygen ions (V-type centres) adjacent to Y^{3+} cations [24]. After annealing and Si-ion irradiation, the shift of SPR to higher wavelengths (~580 nm) was observed (Figure 2b) for the two higher ion fluences. It was connected to Au NPs clustering or increasing the size of NPs [25]. For the lowest ion fluence, the SPR peak was observed at ~530 nm (Figure 2b – turquoise line), which correspond to Au NPs formation after annealing. Further analysis is needed to reach satisfactory conclusions about the size and distributions of Au NPs.

4 Conclusions

The YSZ samples were implanted with 1-MeV Au^+ ions over a broad ion-implantation fluence range at RT. These samples were characterised by RBS, RBS-C, and optical measurement. The combination of these analytical methods provides advantages for a better understanding of structural modifications upon implantation and subsequent annealing and Si^{3+} ions irradiation. The study shows that the formation of Au NPs occurs before annealing especially for ion fluence from 5.0×10^{16} cm^{-2} and that annealing and Si irradiation cause the formation of Au NPs even for the lowest Au fluence (1.5×10^{16} cm^{-2}).

The research has been carried out at the CANAM (Centre of Accelerators and Nuclear Analytical Methods) infrastructure LM 2015056. This publication has been supported by OP RDE, MEYS, Czech Republic, under the project CANAM OP, CZ.02.1.01/0.0/0.0/16_013/0001812.

References

1. G. Sirinakis, R. Siddique, K.A. Dunn, H. Efstathiadis, M.A. Carpenter, A.E. Kaloyeros, L. Sun, J. Mater. Res. 20, 3320–3328 (2005)
2. U. Kreibig, M. Vollmer, Optical Properties of Metal Clusters, Springer Berlin Heidelberg, Berlin, Heidelberg, 1995.
3. L. Montaño-Priede, O. Peña-Rodríguez, A. Rivera, A. Guerrero-Martínez, U. Pal, Nanoscale. 8, 14836–14845 (2016)
4. A.A. Leino, F. Djurabekova, K. Nordlund, Eur. Phys. J. B. 87, 242 (2014)
5. J.J. Penninkhof, T. van Dillen, S. Roorda, C. Graf, A. van Blaaderen, A.M. Vredenberg, A. Polman, Nuclear Instruments and Methods in Physics Research Section B: Beam Interactions with Materials and Atoms. 242, 523–529 (2006)
6. E.A. Dawi, A.M. Vredenberg, G. Rizza, M. Toulemonde, Nanotechnology. 22, 215607 (2011)
7. C.H. Kerboua, J.-M. Lamarre, M. Chicoine, L. Martinu, S. Roorda, Thin Solid Films. 527, 186–192 (2013)
8. F. Ren, X.H. Xiao, G.X. Cai, J.B. Wang, C.Z. Jiang, Appl. Phys. A. 96, 317–325 (2009)
9. P. Wright, Current Opinion in Solid State and Materials Science. 4, 255–265 (1999)
10. S. Park, J.M. Vohs, R.J. Gorte, Nature. 404, 265–267 (2000)
11. Y. Saito, Y. Imamura, A. Kitahara, Nuclear Instruments and Methods in Physics Research Section B: Beam Interactions with Materials and Atoms. 206, 272–276 (2003)
12. N. Docquier, S. Candel, Combustion control and sensors: a review, Progress in Energy and Combustion Science. 28, 107–150 (2002)
13. P.H. Rogers, G. Sirinakis, M.A. Carpenter, J. Phys. Chem. C. 112, 8784–8790 (2008)

14. R. Mikšová, A. Jagerová, P. Malinský, P. Harcuba, J. Veselý, V. Holý, U. Kentsch, A. Macková, Vacuum. 184, 109773 (2021)
15. M. Mayer, SIMNRA, a simulation program for the analysis of NRA, RBS and ERDA, in: AIP Conference Proceedings, AIP, Denton, Texas (USA), 1999: pp. 541–544.
16. J.F. Ziegler, M.D. Ziegler, J.P. Biersack, SRIM – The stopping and range of ions in matter, Nuclear Instruments and Methods in Physics Research Section B: Beam Interactions with Materials and Atoms. 268, 1818–1823 (2010)
17. A. Macková, P. Malinský, A. Jagerová, R. Mikšová, P. Nekvindová, J. Cajzl, R. Böttger, S. Akhmadaliev, Vacuum. 169, 108892 (2019)
18. G. Bohm, G. Zech, Introduction to statistics and data analysis for physicists, Verl. Dt. Elektronen-Synchrotron, Hamburg, 2010.
19. R. Mikšová, A. Macková, A. Jagerová, P. Malinský, P. Slepička, V. Švorčík, Applied Surface Science. 458, 722–733 (2018)
20. H. Paul, Nuclear stopping power and its impact on the determination of electronic stopping power, in: Ft. Worth, TX, USA, 2013: pp. 309–313.
21. X. Hong, S. Xu, X. Wang, D. Wang, S. Li, B.A. Goodman, W. Deng, Journal of Luminescence. 231, 117766 (2021)
22. X. Huang, M.A. El-Sayed, Journal of Advanced Research. 1, 13–28 (2010)
23. V. Morandi, F. Marabelli, V. Amendola, M. Meneghetti, D. Comoretto, Adv. Funct. Mater. 17, 2779–2786 (2007)
24. S. Zhu, X.T. Zu, X. Xiang, Z.G. Wang, L.M. Wang, R.C. Ewing, Nuclear Instruments and Methods in Physics Research Section B: Beam Interactions with Materials and Atoms. 206, 1092–1096 (2003)
25. S. Link, M.A. El-Sayed, J. Phys. Chem. B. 103, 4212–4217 (1999)

High-contrast low-dose proton radiography of thin samples at the Tandetron accelerator

Vaclav Olsansky[1,2*], *Carlos* Granja[3], *Cristina* Oancea[3], *Pavel* Krist[1], *Anna* Mackova[1,4], *Vladimir* Havranek[1] and *Jiri* Bila[2]

[1]Nuclear Physics Institute Czech Ac. of Sc., p.r.i., Husinec-Rez 130, 250 68 Rez near Prague, Czech Republic
[2]Faculty of Mechanical Engineering, Czech Tech. Univ. in Prague, Technicka 4, 160 00 Prague 6, Czech Republic
[3]Advacam, U Pergamenky 12, 170 00 Prague 7, Czech Republic
[4]Faculty of Science, J. E. Purkinje University, Pasteurova 1, 400 96 Usti nad Labem, Czech Republic

Abstract. We investigate the ability of using high-resolution position-sensitive pixel detector and standard non-scanning beams of low-energy protons in air as a flexible tool and simplified technique for density-sensitivity imaging of thin samples. Proton radiography can provide high contrast and low radiation dose delivered to the inspected sample. Density-sensitive contrast response can be provided by a single proton per imaging pixel. For this purpose, we use the silicon semiconductor high-resolution pixel detector Timepix3 to evaluate particle radiography of thin samples with monoenergetic low-energy proton beams from the Tandetron light-ion accelerator. Measurements were performed with various well-defined thin samples. A proton micro beam was used to test and evaluate the technique. Spatial information of the samples is provided by the imaging detector. Density-sensitive contrast is obtained from the measurement of small differences in the deposited energy of transmitted protons across the sample. The transmitted protons are detected with high spatial resolution in the pixel detector. The single particle tracks registered in the detector are analysed by detailed pattern recognition algorithms. Various of these track parameters of spectral response are used for imaging contrast. Resulting proton radiographies for various well-defined thin samples are presented.

1 Introduction and Motivation

Use of protons for imaging of objects dates back to the 1960s, when proton radiography was introduced [1]. Depending on the particle energy and the resulting range penetration in the inspected samples together with experimental setup restraints (e.g., vacuum or in-air measurements), several techniques have been developed according to the capabilities provided by delivered particle beams and available radiation detectors [2]. Scanning transmission ion microscopy (STIM) achieved with high-spatial resolution microbeams and

* Corresponding author: olsansky@ujf.cas.cz

conventional single-pad semiconductor diodes measures the deposited energy of transmitted protons and ions through the imaged sample [3-4]. The single-particle sensitivity and deposited energy response give the capability to provide enhanced resolving power of low-contrast structures [3-5]. On the contrary, in conventional radiation imaging, such as in standard X-ray radiography, changes in beam intensity are used for generation of the contrast mechanism [2]. The use of position-sensitive high-sensitivity imaging detectors enables to use the technique with standard non-focused and non-scanning beams which can be more readily accessed including measurements in air. Even a standard radionuclide alpha source, e.g., ^{241}Am, can be used [5]. In addition, very few particles, even only one proton per image bin can be sufficient to provide contrast information [6]. A benefit of this approach is that the particle goes through the sample, transferring only a small amount of energy yielding a low fluency of the radiation required to provide imaging information. The image sample can be thus exposed to a lower radiation dose than X-rays, which otherwise require irradiation of large number of quanta per imaging bin many of which are absorbed in the sample.

In this work we develop and evaluate a simple and practical approach with low-energy protons in air using high-resolution semiconductor pixel detectors. The goal is to implement the application of high-sensitivity low energy proton radiography at the Tandetron 4130 MC accelerator at the NPI CAS [7-8].

We use of the new Timepix3 ASIC chip detector [9] for enhanced-contrast density-sensitive radiography of thin samples irradiated in air with low-energy protons. A micro beam is used to test and evaluate of the technique. The imaging setup is used direct in-line transmitted geometry. The proof-of-principle and method of image contrast are based on high-resolution analysis of single particle tracks in the pixel detector [10]. We examine the capability to produce contrast images from detailed pattern recognition and spectral tracking morphology analysis of the single particle tracks depending on the deposited energy, size (number of pixels) and energy height of the pixelated signals [11]. Spatial maps can be thus produced of particle energy loss across the sample [10] with high-resolution, enhanced contrast and low dose. Radiographies of event rate are also produced which serve for derivation of total fluency and deposited dose.

2 Methods and materials

2.1 Pixel detector Timepix3, deposited energy contrast imaging

We use the pixel detector Timepix3 [9] equipped with a 500 μm Si sensor. The detector provides a high-granularity matrix of energy-sensitive pixels (256 × 256 pixels). The pixel pitch size is 55 μm for a total sensitive area of 14.08 mm × 14.08 mm = 1.98 cm2. The detector is operated and readout with the fast data-rate AdvaPIX interface. Data are readout in continuous pixel stream data driven mode including acquisition visualization online with the PIXET software tool. The detector sensor surface has a thin non-sensitive is coated layer (about 1 μm, Aluminium) which makes the sensor also non-sensitive to light.

The principle of density-sensitive contrast imaging is based on the precise measurement of small changes in the proton deposited energy [3-6, 10]. This method requires a well-defined (monoenergetic, collimated/scanning, parallel) beam together with a spectral sensitive position-sensitive detector [5]. The initial energy of the accelerated protons (2.9 MeV in this work) is known and changes of proton deposited energy after passing sample are accurately measured by the pixel detector. Each particle makes a characteristic trace in the pixel detector consisting of various pixels called a cluster [5, 11]. The micro-scale cluster morphology and spectral response of the track is correlated with the deposited energy [11]. Raw data are converted to energy using a per-pixel calibration [12]. The transmitted protons are registered in high spatial resolution by the detector with sub-pixel (detector pixel pitch size 55 μm) resolution [5]. Position resolution up to few μm level is made possible by pattern recognition and charge sharing model fitting [5]. The imaging contrast principle requires few particles per imaging bin.

Contrast imaging can be derived from spectral-tracking parameters of the registered pixelated clusters [10]. Several of these pattern-recognition parameters are correlated with the deposited energy. The parameters which we examine in this work are: deposited energy, cluster area size and cluster per-pixel height of single particle track. Event counts of localized protons in the given imaging bin are also evaluated. The deposited energy is given by the sum of all per-pixel energy values of the cluster. Cluster area size is the number of cluster pixels; cluster per-pixel height corresponds to the maximum energy value of the pixels in the given cluster.

2.2 Measuring setup, geometry

Measurements were performed at the Tandetron 4130 MC accelerator. The pixel detector was scanned in air with a proton micro beam of energy 2.9 MeV. The detector and sample were placed in air perpendicular to the beam axis. The distance between the accelerator beam nozzle and the detector sensor was 7 mm, (the setup is shown in Fig. 1).

The imaged samples were held by plastic frames of larger size. The samples on the frames were mounted in close geometry in front of the detector. The detector and sample were mounted on a movable sample holder, which was used for spatial scanning of large samples. Small homogeneous samples were irradiated using the automatic beam scanning system, which covered a square field of 1.5×1.5 mm^2.

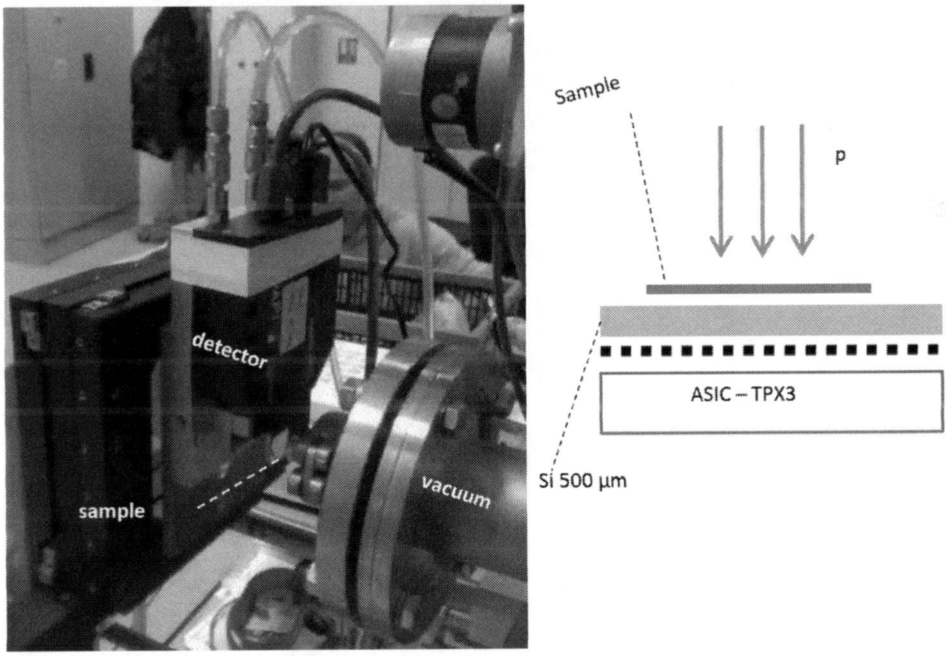

Fig. 1. Photo (left) and scheme (right) of the imaging setup in air. The detector is placed at a distance of 7 mm from the accelerator beam nozzle. The deposited energy and position of the transmitted protons are registered in high-resolution by a Timepix3 detector with a 500 μm silicon sensor (shown in green).

Table 1. Imaging samples used. Composition, labels, foil thickness are given.

Label	Sample, composition and size	Composition, layout
A	No sample	Air
B	Mylar foil 5 μm	Homogenous sample, one foil
C	Al foil 5 μm	Homogenous sample, one foil
D	Stairs of Al foils	Stacked sample, stairs of foils with total thicknesses 0, 5, 10, 15, 25 and 30 μm
E	Stairs of Mylar foils	Stacked sample, stairs of 5 μm foils, with total thicknesses from 0 to 20 μm
F	Edge of metallic Mylar foil	Edge of the foil is in the middle of a scanned area, the metallic layer starts approximately 1 mm next to the edge of foil
G_1	Metal grid 0.25 mm	Grid placed directly on the exit window
G_2		Same grid, rotated at other angle

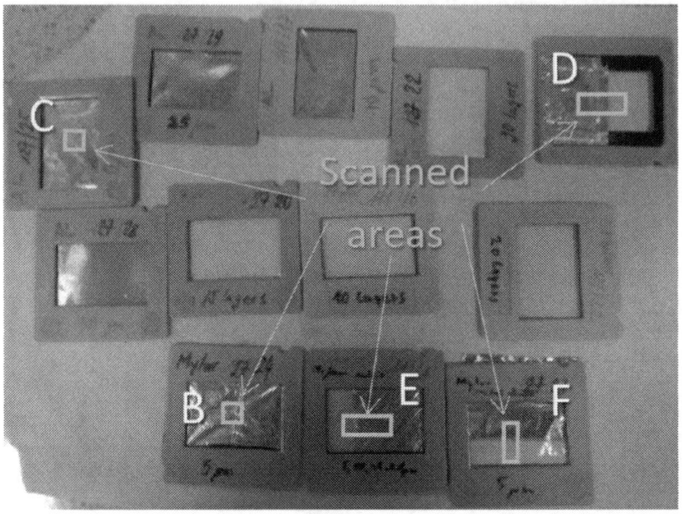

Fig. 2. Imaging samples used: simple Mylar foil 5 μm (B); simple Aluminium foil 5 μm (C); stair of Aluminium foil (0, 5, 10, 15, 25 and 30 μm) (D); stairs of Mylar foil (0, 5, 10, 15 and 20 μm) (E); edge of 5μm Mylar foil (F). Orange rectangles and squares represent the scanned areas.

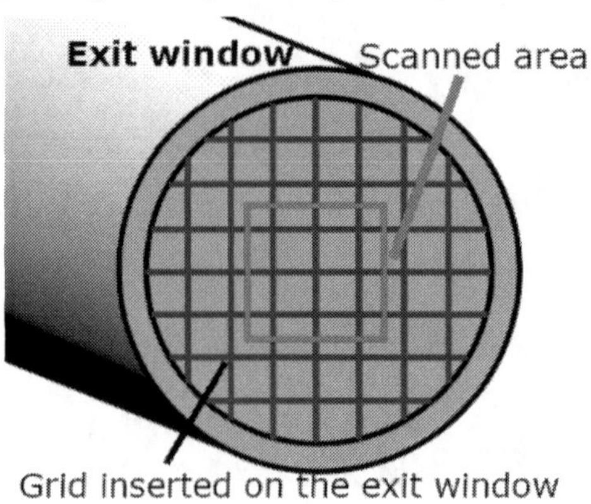

Fig. 3. Scheme of the metallic grid sample (G_1 and G_2) placed closely on the exit window. The beam scanned area is illustrated (red square).

2.3 Imaged samples

Well-defined samples were assembled from layers of thin Mylar and Aluminium foils as well as a thin metallic grid to serve as evaluating imaging samples. The detector was also irradiated without sample in order to examine the beam spectral and beam size profile homogeneity. The sample list is given in Table 1. Several samples are shown in Fig. 2. The grid sample is illustrated in Fig. 3.

3 Measurements and image processing

3.1 Data acquisition, data processing

Samples were imaged with a low intensity beam (flux $\approx 10^3$ protons/cm^2/s). A low intensity beam ($< 10^{-13}$ A/cm^2) is required for detailed event-by-event detection in the pixel detector [11]. This acquisition mode allows registering and analysing in detail the single particle tracks for precise high-resolution spectrometry (deposited energy). Collected raw data are processed offline starting with the application of the per-pixel calibration and pattern recognition and identification of pixelated clusters. Detailed event lists are produced with spectral, tracking, position and time information. For each event the position in X and Y fitted coordinates of the cluster centroid is determined with sub-pixel μm level resolution [4]. Also deposited energy, maximum per-pixel energy value (height) of the cluster, number of pixels (area) and further spectral-tracking pattern recognition parameters such as cluster roundness and linearity are registered [11].

Radiographies can be produced in the form of spatial distribution of these parameters including the event count radiograph. The image spatial bin of the given radiograph can be scaled in proportion to the event fluency to e.g., the detector pixel size i.e., 55 μm, or smaller e.g., 10 μm [9]. The information produced is namely the material density map of the imaged sample. The sensitivity and dynamic range of the image contrast information are given by the value and extent of the selected cluster parameter. If multiple events of cluster centroids located in the same image bin, the average or maximum value are evaluated and displayed. Proton radiographies have been made for all samples using the selected parameters (counts, deposited energy, cluster area and heights). For these quantities the average or the maximum value are evaluated. Resulting radiographies can be produced from all events i.e., all clusters or from filtered data. Data are filtered in dependence to the individual parameters and their combinations.

3.2 Proton radiographies

Radiographies of the 5 μm Al foil stack (sample D) imaged by a 2.9 MeV proton scanning micro beam is shown in Fig. 4. The image shown is produced with spatial bin equal to the detector pixel size (55 μm). In the data a basic filter was applied (clusters greater than 7 px size), which corresponds to protons both primary beam as well as scattered and beam halo protons. This filter removes non-proton events, such as X rays and unwanted or background electrons. For the same data set are shown four radiographies of different contrast response each produced by evaluation of different cluster parameter. Information is derived on the energy loss of the probe particle and material composition density map across the imaged sample. The image exhibits a main central component, resulting from the primary beam non-scattered particles in the scanning microbeam. In the periphery there is a small component of decreasing radial intensity. Consisting of scattered particles this periphery halo component can be resolved also by the deposited energy which is nearly half that of non-scattered primary beam particles (see Fig. 4a, c). The evaluation and quantification of this and further effects observed are subject of future work.

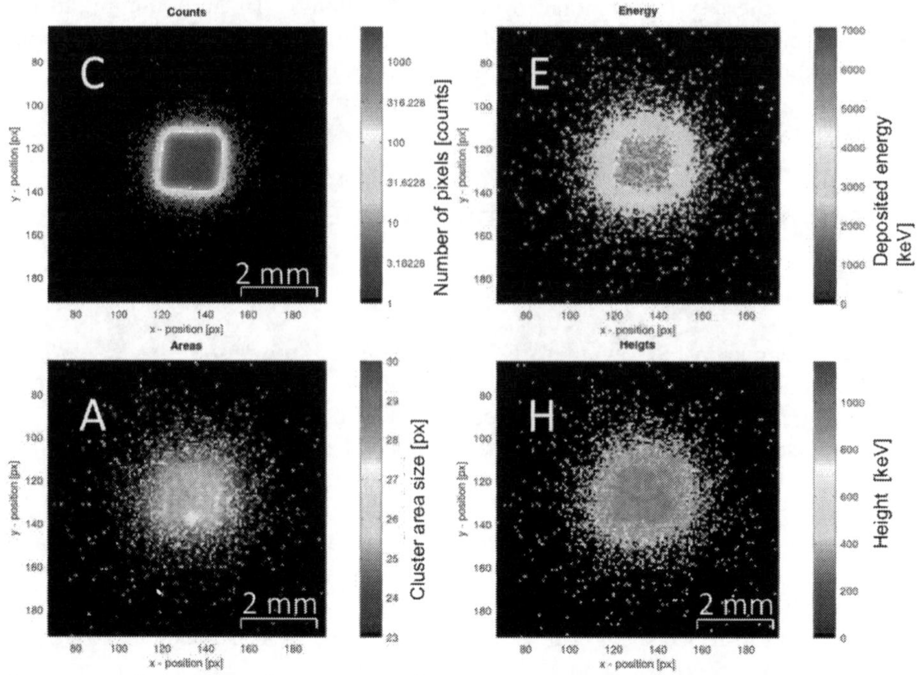

Fig. 4. Proton radiographies of sample C with 2.9 MeV protons measured by a Timepix3 500 μm silicon sensor detector. The images are based on event count (C), deposited energy (E), cluster area sizes (A) and cluster height (H). All images are made from the same set of filtered data (filter used: A > 7 px). The spatial bin is equal to the detector pixel size (55 μm).

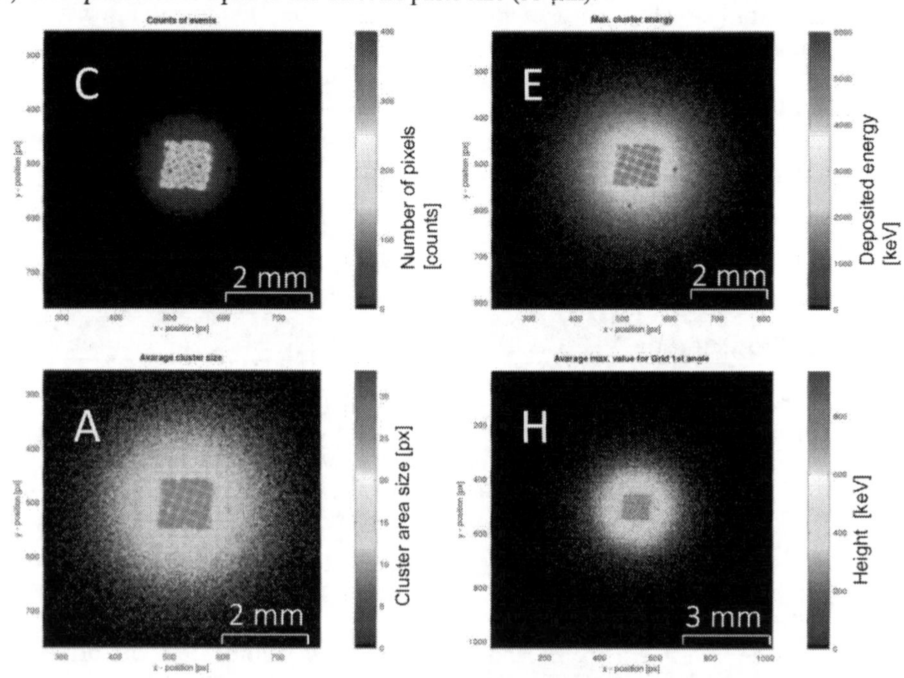

Fig. 5. Similar to Fig. 4 showing proton radiographies of sample G_1. Images are made from all data (no filtering applied). Image produced with subpixel resolution (one spatial bin corresponds to 13.75 μm).

Depending on the total event fluency, i.e., total number of particles registered, it is possible to image radiographies with varying spatial resolution. A radiograph with subpixel resolution is shown in Fig. 5 of the grid sample (sample G1) whose scheme is illustrated in Fig. 3. Four radiographies of different contrast response are shown. In the images presented, no event filter was applied.

Radiographies with enhanced contrast can be produced by combining filtered data with non-filtered data i.e., the image of all cluster events, see Fig. 6. Structures of the same material and small thickness down to the μm range can be resolved.

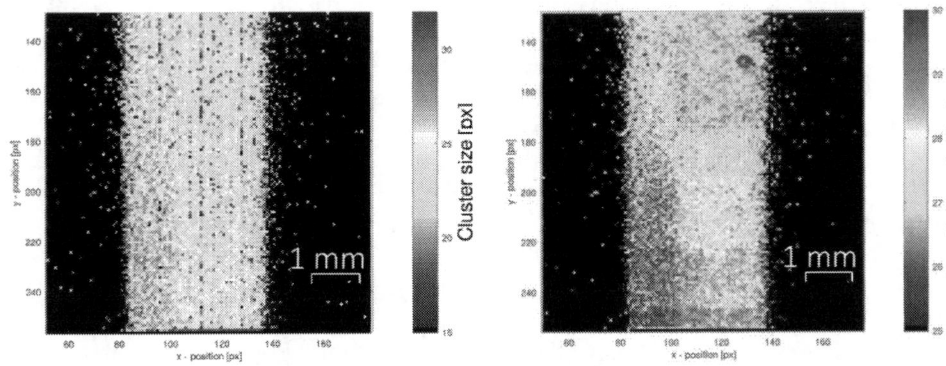

Fig. 6. Proton radiographies produced from non-filtered i.e., all data (left) and filtered data (right). The sample E is imaged with 2.9 MeV protons. The right filtered image is made from events selected in the following range: Deposited energy from 0.8 to 3.6 MeV, cluster area greater than 10 px and cluster energy height greater than 300 keV. Radiographies shown in spatial bin of one detector pixel i.e., 55 μm.

3.3 Evaluation of the Technique

This technique sensitivity allows resolving materials of thickness at the μm level in wide dynamic range limited by the range of the charged particle used across the sample. The maximum thickness size of imaged samples with 2.9 MeV protons is several tens of μm depending on the material density. The contrast principle of this technique allows even just one particle to be required per spatial bin. Depending on the flux used ($< 10^5$ particles/cm^2/s), selected spatial resolution (achieved down to the few μm level) and overall image sample area size, complete scans can be performed in few seconds up to few minutes. The counting statistics which depend on the chosen spatial resolution are summarized in Table 2. Values given for the lowest statistics (i.e., on average one particle per spatial bin). Estimated typical values of total dose given to the sample are included. These values are calculated for a mean deposited dose in the sample of 200 pGy per proton.

Table 2. Lowest event count statistics depending on the spatial resolution. Values given for a minimal event rate of 1 proton per imaging bin (spatial resolution) for a sample of size 10 mm × 10 mm. Scanning time given for a particle flux of 1×10^4 protons/cm^2/s. Total dose values are included.

Number of particles	Spatial bin size	Scanning time	Total dose
3.3×10^4	1 bin ≅ 1 px ≅ 55 μm	3.3 s	6.6 μGy
1×10^6	1 bin ≅ 10 μm	100 s	200 μGy
4×10^6	1 bin ≅ 5 μm	400 s	800 μGy

4 Conclusions

Use of a spectral and imaging detector such as Timepix3 enables to apply the technique of deposited energy sensitive radiography using standard proton beams in air. Even a standard radionuclide alpha source can be used (limited contrast, limited count rate). We evaluated and tested the detector resolving power on well-defined thin samples using a mono-energetic collimated proton micro beam. We examined the image contrast response on various parameters of the single particle tracks in the pixel detector. These can be combined to refine further the image contrast and provide information on the sample material composition (subject of future work). The spatial resolution of the resulting radiographs is in the range 55 μm (detector pixel size) to 5 μm (limit by the detector resolution and beam straggling in air) depending also on the event rate (total particle fluency). Thanks to the detector high sensitivity the event rate per imaged spatial bin can be even one particle per pixel. Structures of similar and even of the same material can be resolved down to small thickness in the μm range. The achieved resolution and single particle sensitivity enable to minimize the radiation dose and resolve composition and details hard to achieve with conventional or other non-destructive techniques. The technique allows to resolve components in the transmitted beam such as scattered particles forming a periphery halo. Quantification and evaluation of distortion of the resulting image are in progress. Future work includes combined evaluation of multi-parametric radiography and further contrast enhancement by data post-processing techniques. The imaging sensitivity can be further improved by measurements in vacuum. Higher sensitivity can be provided by use of heavier particles such as ^{12}C ions.

The authors appreciate the remarks and valuable suggestions given by the reviewers. Measurements were carried out at the CANAM infrastructure of the NPI CAS Rez supported through MŠMT project No. LM2015056.

References

1. A. M. Koehler, Science **160** (1968)
2. R.P. Johnson, Rep. Prog. Phys. **81**, 1 (2018)
3. I.F. van Maanen, et. al., NIM-B **113** , 382-386 (1996)
4. R. Ortega, et. al., NIM-B **181**, 475-479 (2001)
5. J. Jakubek, A. Cejnarova, T. Holy et al., NIM-A **591**, 155-158 (2008)
6. J. Jakubek, JINST **4,** P03013 (2009)
7. O. Romanenko, V. Havranek, A. Mackova et al., Rev Sci Instrum. **90**, 013701 (2019)
8. A. Mackova, P. Malinsky, M. Cutroneo et al., Eur. Phys. J.+ **136**, 5, 558 (2021)
9. T. Poikela, J. Plosila, T Westerlind, et al., JINST **9**, C05013 (2014)
10. V. Olsansky, C. Granja, C. Oancea et al., JINST (2021) *submitted*
11. C. Granja, J. Jakubek, S, Polansky et al., NIM-A **908**, 60-71 (2018)
12. J. Jakubek, NIM-A, **633**, S262-S266 (2011)

Enhanced thin film analysis via High Resolution RBS using the NEC CARBS system

Thomas Pollock[1*], *Eric* Alderson[1], *Kevin* Berk[1], *Levon* McQuown[1], *Ilya* Kostanovskiy[2], *Peng* Wang[2], *Dmitry* Knyazev[2] and *Stuart S. P.* Parkin[2]

[1]National Electrostatics Corp. 7540 Graber Rd, Middleton, WI 53562, USA

[2]Max Planck Institute of Microstructure Physics, Weinberg 2, D-06120, Halle Saale, Germany

Abstract. Advancements in thin film deposition techniques can now produce films of only a few monolayers in thicknesses, with multiple applications emerging in the nano technology field. This maturing manufacturing technique is driving the need for diagnostics tools able to accurately measure depth profiles. To meet this need, the Compact Automated Rutherford Back-Scattering (CARBS) system is under development at National Electrostatics Corp. (NEC) for nanometer thin film analysis using High-resolution RBS (HRBS) within a 4 x 4meter footprint. We present the recent development of the system and demonstrate a study of HRBS applied to a 30nm CoAl alloy film. We compare the performance of the CARBS system with the conventional NEC HRBS end station and discuss the advantages of HRBS over SIMS method.

The advancements in production of monolayer level films [1] require diagnostic tools with accuracy of a few Angstroms. The several MeV beam energies typically applied in Rutherford Backscattering Spectroscopy (RBS) result in the penetration depth of up to a few micrometers. As a result, the effective depth resolution is in the range of a few tens of nanometers [2,3]. This makes RBS analysis not sensitive enough to resolve the fine structure within a film. Figure 1 demonstrates a RBS spectrum measured with 1.9MeV He+ beam in the RBS Facilities at MPI of Microstructure Physics. A sputter-deposited film made at MPI of Microstructure Physics consisting of 3nm MgO capping, 30nm CoAl alloy on the MgO substrate is taken as a demo system. To remove the charging due to insulating MgO capping an additional 30 nm thick carbon layer was extra deposited before RBS measurement. The RBS spectrum measured with 169 degrees and 90 degrees scattering angle allows to resolve Co and Al peaks and provides an accurate measure for the Co-Al elemental ratio within a film but gives no information on the homogeneity of the film. Obviously, the beam energy has to be reduced to get a more detailed shape of Co and Al peaks. The RBS spectrum measured with He+ beam 400keV is shown in Figure 1 and provides an improved depth profile and reveals an important concentration gradient within the film (Figure 1 – middle). The further improvement in the depth resolution is limited by the energy resolution of the standard Si implanted detectors (Ametek, 15 keV). To achieve the desired energy and depth resolution the High-Resolution RBS technique has been developed in the last decades [4]. This technique utilizes a magnetic spectrometer with MCP detector and allows to reduce the energy resolution up to 1 keV.

*Corresponding author: pollock@pelletron.com

One of the first dedicated HRBS systems is the NEC CARBS. Here we present the recent improvements in the CARBS system and discuss the performance as compared with the standard NEC HRBS end station. We show HRBS analysis of the introduced CoAl sample and discuss the improvement in the performance. Finally, we compare the HRBS sensitivity for depth profiling with those of SIMS method.

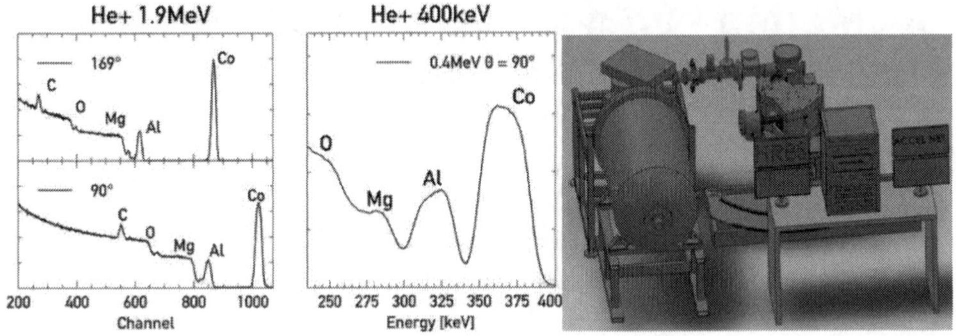

Fig. 1. RBS spectra of the test sample (MgO/30nm CoAl/3nmMgO/30nm Carbon) as measured at different scattering angles θ with the 1.9MeV (left) and 400keV (middle) He+ beam using Si implanted detector with 15keV energy resolution. Right - overview of the NEC CARBS system

The NEC CARBS system is shown in figure 1 (right). The 0.5MV Pelletron® accelerator is equipped with the RF source to deliver 1mm in diameter He+ and H+ beams of 50 to 100nA on target. Mass and energy analysis is accomplished with a 90° magnet followed by beam size measurement, definition, and diagnostics. The analysis chamber has the base pressure of 1.3E-5 Pa and allows for sample positioning with a 5-axis manipulator. The high-resolution magnet separates the scattered beam for acquisition with the detector. The latter is a 100 x 15mm micro channel plate (MCP) with resistive anode encoder (RAE) with ~50um spatial resolution in the horizontal plane. Data collection and system operation are automatically completed by the HRBS and AccelNET computers. Scattering angles between 45° and 125° are available without breaking the vacuum. Additionally, a standard RBS detector is available and can be positioned remotely.

The requirement to meet in the HRBS is the best possible energy resolution. In conflict with improving the energy resolution is obtaining a reasonable detector solid angle. This is due to increasing kinematic energy spread in the scattered beam with increasing solid angle. Typical HRBS systems have a solid angle of 0.4mStr. This results in an angular spread of ±0.7° which induces a kinematic energy spread. The energy spread becomes greater than ±1keV for Silicon and lighter elements and is a maximum for H in Forward Recoil Spectroscopy affecting the final energy resolution. To minimize this effect of the energy spreading due to the solid angle effects we performed OPRA/TOSCA [4] simulations and redesign the new magnet poles. From the magnet simulation, the focal points of the scattered He+ are perpendicular to the central ion trajectory and used for correct detector placement see Figure 2 (right), and optimize the beam focusing at the MCP detector. The new magnetic poles [5] allow to optimize the focus of the incoming particles on the one-dimensional detectors such that the scattered particles of the same energy are perfectly aligned vertically at the detector. Moreover, with the optimized detector placement the total energy width is ~28% of the central ray energy (~22% previously). To take advantage of the energy resolution of the magnet the MCP/RAE detector pulse processing electronics time resolution is set for ~10ns [6]. Figure 2 (left) confirms the ultimate 1keV HRBS energy resolution of the CARBS system as measured with the HfO_2 film on Si [7].

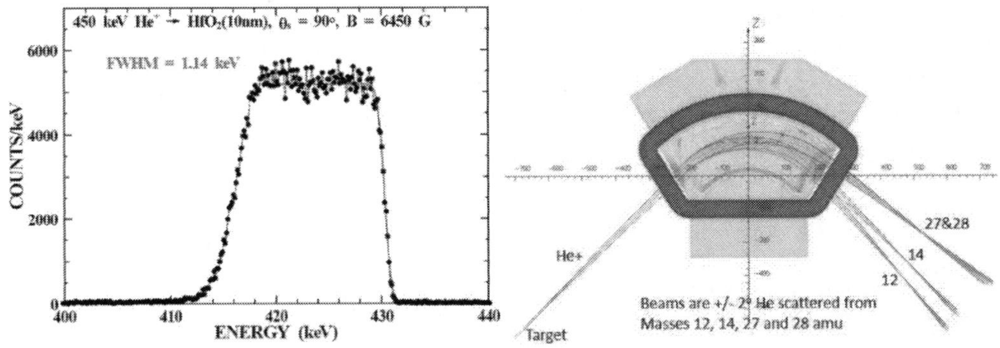

Fig. 2. Right: Simulation of the magnetic Focusing of He+ scattered from indicated masses. Left: Hf edge of the 10nm HfO2 film on the Si substrate as measured with HRBS. The edge width of 1.14keV defines the ultimate energy resolution of the CARBS system

We have performed the comparison of the HRBS performance between the standard NEC RC43 end station at the Max Planck Institute of Microstructure Physics and the new CARBS endstation using an introduced test sample (MgO substrate/ 345ACoAl/ 35A MgO/308A Carbon). Figure 3 (top) shows the HRBS spectra around Co and Al peaks as measured by the two systems. The resulting profile allows the detection of the concentration gradient within the film. The elemental ratio is estimated using SimNRA code and is varying from Co 0.68 Al 1 on the surface towards Co 0.73 Al 1 in the middle of the film and Co 0.81 Al 1 on the interface with substrate. A classical RBS analysis applied for that sample provides only the averaged composition of Co 0.75 Al 1. Although the shape of the peak measured by the two systems is similar, the improvements in the CARBS EndStation allows it to achieve a broader energy window for the same parameters as seen from the broader line shapes.

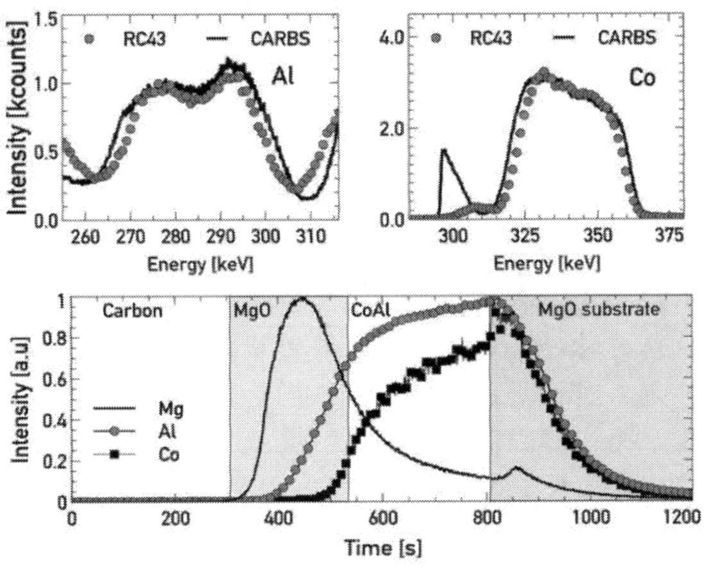

Fig. 3. Top - HRBS data of the Co and Al peaks as measured with RC43 and CARBS systems. Bottom - SIMS data as measured with 3keV Xe+ beam

For comparison, we performed the SIMS analysis of the same sample using a 3keV Xe+ beam. The SIMS depth profile is shown in Figure 3 (bottom). Different layers of the sample are marked for clear navigation. The intensities go down for the etching time above 800s as sputtering of the insulating MgO substrate slows down significantly due to charging effects. Compared to the HRBS data the accurate evaluation of the elemental stoichiometry within the CoAl film (time range 500 - 800 s) is not feasible.

To conclude, we have presented a comparisional demonstration of the recent improvements to the NEC CARBS System for HRBS. The development of the magnetic poles in the detector using accurate numerical simulations and updated electronics with higher time resolution allows to achieve very high energy resolution of 1 keV, an increased energy window while remaining in the compact layout. We applied HRBS analysis to study a thin CoAl film for accurate depth profiling and show the advantages of CARBS system as compared to conventional NEC RC43 end station. Finally, we discuss the advantages of HRBS method as compared to SIMS.

Literature

1. A. Jain. Hafnium Dioxide (HfO2) High-k Dielectric thin film formation for Gate insulators in IC's (2003)
2. J. Perkins et al. New high performance transparent conducting oxides, NREL/CP-520-37081 (2005)
3. S. Bauers et al. Epitaxial growth of rock salt $MgZrN_2$ semiconductors on MgO and GaN, Appl. Phys. Lett. 116, 102102 (2020)
4. K. Kimura, AIP Conference Proceedings 475, 500 (1999)
5. N. White, Private Communications, 2021.
6. J. Wiza, Microchannel Plate Detectors, Nucl. Instr. And Meth. 162 (1979) 587.
7. K. Kimura, Private Communications, 2021.

Energy Sensitive Imaging of Focused and Scanning Ion Microbeams with μm Spatial and μs Time Resolution

Carlos Granja[1*], *Cristina* Oancea[1], *Anna* Mackova[2,3], *Vladimir* Havranek[2], and *Vaclav* Olsansky[2,4]

[1]Advacam, U Pergamenky 12, 170 00 Prague 7, Czech Republic

[2]Nuclear Physics Institute, Czech Academy of Sciences, Husinec 130, 250 68 Rez, Czech Republic

[3]Faculty of Science, J. E. Purkinje University, Pasteurova 1, 400 96 Usti nad Labem, Czech Republic

[4]Faculty of Mechanical Engineering, Czech Technical University in Prague, Technicka 4, 160 00 Prague 6, Czech Republic

Abstract. We inspected and imaged the delivery of ion microbeams with spatial, time and energy sensitivity. Quantum imaging registration event- by-event is provided in high spatial and time resolution with the position- sensitive semiconductor pixel detector Timepix. The detector is operated as a miniaturized radiation camera for flexible measurements at room temperature and in vacuum. Imaging information on beam profile, spatial and time distribution, flux, homogeneity, and deposited energy for individual beam particles is provided. Focused and scanning beams can be imaged and evaluated online. Single particles are registered by the detector including spectral (deposited energy) information on their position at the μm and μs level. Delivered beams can be characterized also in terms of composition by resolving background and unwanted components such as electrons and X rays from primary beam particles. Ion groups of different energy including doublets or scattered particles can be identified. The technique is applicable for ions of energy above few hundred keV and beams of low intensity, below 10^5 particles/cm^2/s.

1 Introduction and motivation

Particle accelerators with advanced beam delivery systems provide highly precise, spatially tailored and time-stamped scanning beams at the micrometer scale [1]. It is desirable to independently verify and monitor with sufficient resolution the delivery of such microbeams. The characterization and spatial mapping of such microbeams would ideally include parameters such as beam size, spatial and time distribution, particle composition, direction and deposited energy. Such requirements and related applications can be found in ion beam analysis (IBA) studies, such as scanning ion transmission microscopy (STIM) [2] and ion beam lithography [3].

* Corresponding author: carlos.granja@advacam.com

These tasks can be addressed by high-granularity imaging detectors such as the hybrid semiconductor pixel detector Timepix [4] which provides spectral and tracking response at the pixel level with high resolution for a wide range of particle types [5-7]. Extensive information and high spatial resolution are achieved by placing the pixel in the particle beam (for low beam intensities) and registering single-particle tracks. Detailed event-by-event processing enables particle identification and spectral tracking response to single particles [7-8]. Such high-resolution mode can be applied on a wide variety of radiation fields with particle flux up to about 10^5 particles/cm^2/s (equivalently, up to 10^{-1} pA beam current intensity). The limit of beam flux at which the detector operates in high-resolution event-by-event detection, is set by the particle type, the particle direction and energy loss in the detector sensor [5]. The Timepix detector operates in frame mode with exposure time for example at the ≈ ms level or preferably longer. This level ensures the detector proper operation and complete collection of the deposited charge especially by high energy loss particles especially ions.

The detector and developed technique can be also applied to the evaluation, imaging and monitoring of ion beams of larger beam size at the mm and cm scale [9-10]. Existing ion-beam monitoring sensors such as single-pad detectors have limited resolving power and narrow field-of-view (FoV) requiring the use of collimators. Track-etch detectors provide high spatial resolution but require offline processing. On the other hand, high-sensitivity pixel detectors with integrated electronics provide online response, high spatial resolution, wide FoV, spectral-tracking response and enhanced sensitivity for precise measurements [11]. The pixel detector implemented in miniaturized readout electronics stands as a compact highly integrated camera [6] which can be mounted without complexity and operated by standard computer.

2 Instrumentation and measurements

2.1 Miniaturized radiation camera MiniPIX TPX

Position-sensitive detection of single particles with energy and time information can be provided with high-resolution by the hybrid semiconductor pixel detector Timepix [4]. The detector consists of the Timepix ASIC chip bump bonded to a semiconductor radiation-sensitive sensor. The Timepix detector operates at room temperature including also measurements in vacuum. The detector was equipped with a 300 μm thick silicon sensor which was operated with a relatively low applied bias (+30 V) in order to suppress the per-pixel high-energy saturation and enhance the charge sharing effect. Timepix provides a high granularity matrix of independent pixels of total sensitive area 1.98 cm^2 (256 rows × 256 columns = 65.536 pixels of total size of sensitive area 14 mm × 14 mm = 1.98 cm^2) – see Figure 1a.

Timepix registers single particles and measures at the pixel level the deposited energy, event count rate or time of interaction. The detector is operated with integrated power and readout electronics provided by the miniaturized radiation camera MiniPix TPX [6] shown in Figure 1b. The device is controlled and readout by single USB 2.0 cable to a standard PC or laptop running the cross-platform software tool PIXET.

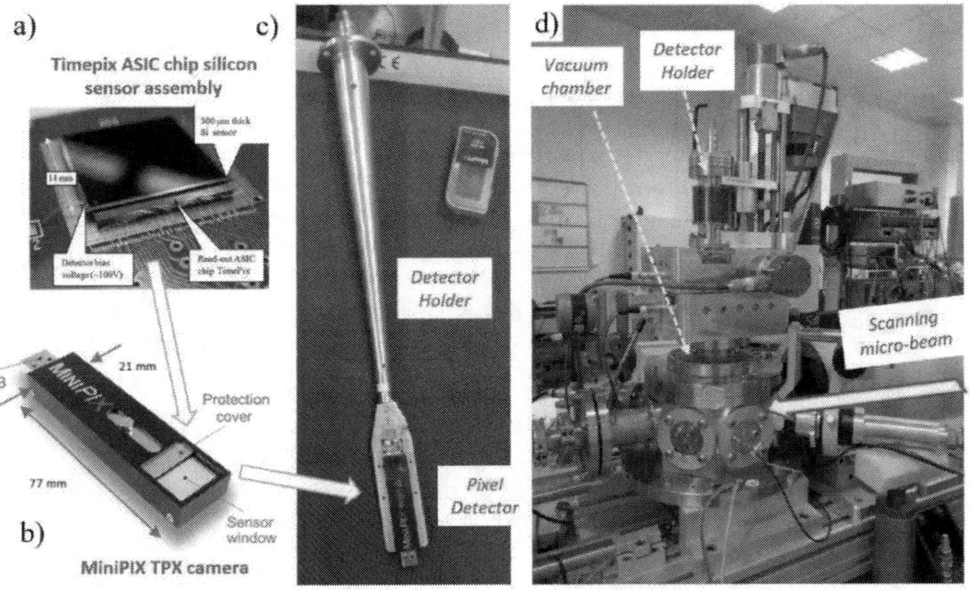

Fig. 1. (a): Pixel detector Timepix ASIC chip – silicon sensor (300 μm thick silicon) assembly. (b) Miniaturized radiation camera MiniPIX TPX operated at room temperature and in vacuum. (c) Detector and holder with thermal contact for passive cooling. (d) Experimental vacuum chamber at the microbeam guide at the Tandetron 4130 MC accelerator, NPI CAS in Rez.

2.2 Ion microbeams at the NPI Tandetron

In this work we used microbeams of ^{12}C ions of two energies (4 MeV and 10 MeV) delivered by the 4130 MC Tandetron accelerator of the NPI CAS Rez [1]. Measurements were performed in vacuum (< 10^{-5} mbar) at the microbeam guide. Focused and scanning microbeams delivered in various modes and configurations in terms of energy, spatial and time distribution were examined and evaluated.

2.3 Measurements

The spatial scanning mode of beam delivery was studied. Adjustable accelerator beam parameters tested included: beam spot size, irradiated beam area, spatial scanning modes and frequencies, beam profile, flux and time distribution. The Timepix detector was operated and readout in frame mode. Depending on the beam flux and event rate at the detector position the frame acquisition time was set in the range from hundreds of μs to few seconds in order to register isolated particle tracks and keep the pixel occupancy low (< 10%). The detector pixels operation mode was used on different measurements on time-over-threshold (ToT) energy mode, designed for precise energy loss registration, or time-of-arrival (ToA) mode, designed for detailed time registration. The detector was operated and readout by standard PC/notebook running Windows using the PIXET software. Raw data were processed offline.

3 Methods and results

3.1 Correlated detection: position, time, deposited energy

The Timepix detector registers single particles in the form of pixelated tracks, called *clusters*, as shown in Figure 2. The charge deposited by charged particles in the center of the cluster is spread out into neighboring pixels due to charge sharing in the semiconductor sensor at room temperature [5,12]. This effect enables to determine, by model fitting of the pixelated cluster, the position of interaction with sub-pixel resolution. For heavy charged particles, such as ^{12}C ions, the spatial resolution can reach the μm level [12]. At the frame level Timepix measures in high resolution for each particle the position and the deposited energy or the position and the time of interaction as shown on top and bottom rows of Figure 2, respectively.

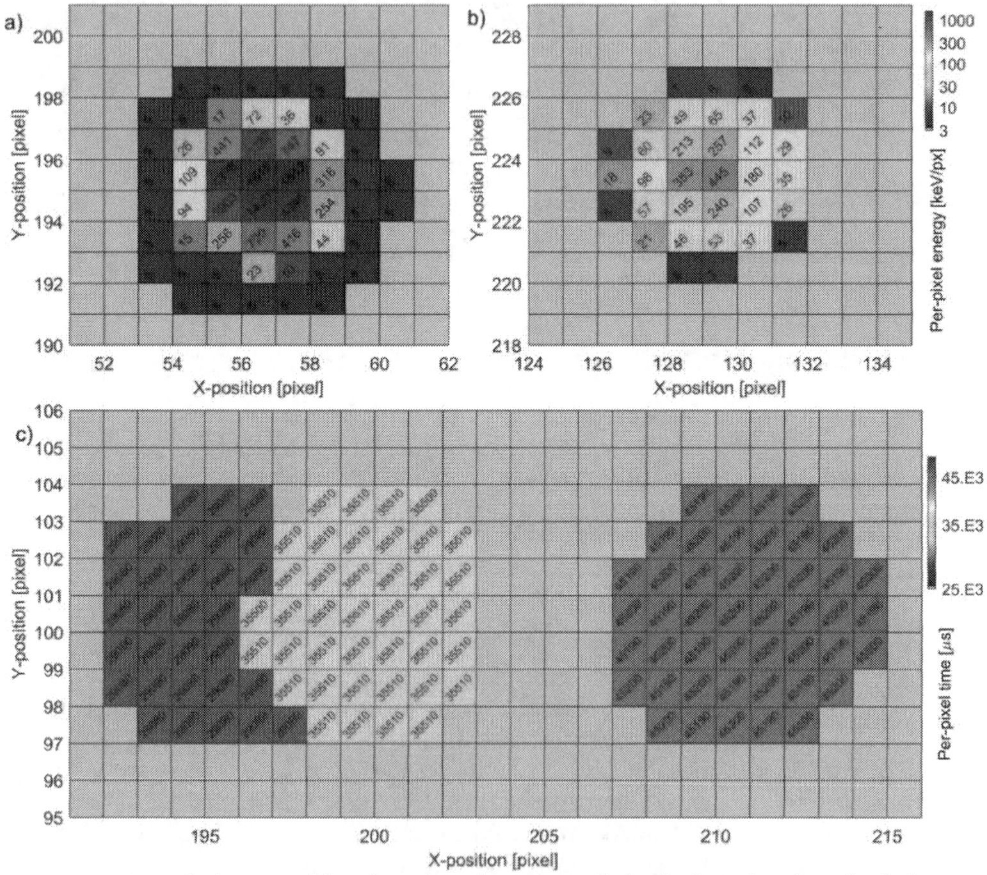

Fig. 2. Spectral- and time-sensitive imaging detection of single ions by the pixel detector Timepix with a 300 μm silicon sensor operated in per-pixel *energy* (top row) and *time* (bottom row) modes. Pixelated tracks, called clusters, are shown for single ^{12}C ions of energy 10 MeV (a) and (c) and 4 MeV (b). The per-pixel energy in keV or time in μs are displayed by the color bar (exact values are typed in the hit-pixels). A small portion of the Timepix detector area is shown – on the top row: 11 × 11 pixels = 605 μm × 605 μm = 0.37 mm^2, and on the bottom row: 11 × 25 pixels = 605 μm × 1375 μm = 0.83 mm^2.

Information on the deposited energy can be recorded in high energy resolution operating the detector in per-pixel (called time-over-threshold ToT) energy mode – see Figure 2, top row, displayed in keV per pixel. The particle deposited energy is given by the sum of the per-pixel energies (shown in color) over the hit pixels in the cluster.

Information on the time of interaction at the pixel level can be measured by operating the detector in time-of-arrival (ToA) mode as shown in Figure 2c. The time-of-interaction registered at the per-pixel level with μs level resolution is displayed by the color bar in Figure 2c. The range and time resolution can be adjusted by varying the value of the Timepix high-frequency clock. Typical values are listed in the Table 1. The extent is given by the per-pixel readout range of 11.810 bins [4]. Each particle registered in the detector is thus assigned a time stamp with resolution at the μs level.

Table 1. Time range and resolution for time-of-arrival (ToA) operation of Timepix. Values are given for selected values of the detector high-frequency clock commonly used.

Timepix clock	Time range	Time resolution
10 MHz	1.8 ms	0.1 μs
1 MHz	11.8 ms	1 μs
100 kHz	118 ms	10 μs
10 kHz	1.18 s	100 μs

Two or more particles can interact in the same spatial region of the detector during the same frame acquisition interval. The resulting overlapping signal in the detector, see doublet event in Figure 2c on left, can be resolved when Timepix is operated in time mode. This setting can be used to examine highly focused and spatially scanning microbeams.

Operating the detector in time mode it is still possible to derive limited information on the particle deposited energy. This is possible by exploiting the spectral sensitivity of the size of the pixelated clusters (i.e., the number of pixels in the cluster). This can be observed in Figure 2 between the registration of ^{12}C ions of 10 MeV and 4 MeV, respectively. By taking into account also the morphology-tracking response of Timepix it can be also derived information on particle-event type [5]. This is possible by cluster morphology and pattern recognition analysis of the characteristic pixelated tracks of Timepix to radiation components in terms of particle types, such as X rays, electrons, protons and ions [5].

3.2 Time-correlated spatial mapping of ion microbeam spot

The time-stamped spatial distribution of single particles from a scanning microbeam of 10 MeV ^{12}C ions is shown in Figure 3. The microbeam was set to a scanning region of spot area of 1mm × 1mm and scanning speed of 4 μm/ms. The beam covered the entire spot region once in nearly 5 min 30 sec during which around 700 particles were registered by the detector in over 11 k Timepix frames of 10 ms acquisition time. The detector live time (113 s) is about one third (34.5 %) of the total beam spot time irradiation (327 s). The detector deadtime of each frame readout was 19 ms. Due to the detector deadtime of frame readout, the detected particles are just about one third of all particles delivered by the microbeam. Particles are also detected beyond the scanning beam region as can be seen in the image map. The incidence and time and spatial characteristics of such events can be evaluated (subject of future work).

The Timepix detector was operated in time mode (see Figure 2c and Figure 3) for registration of single particles with high timing resolution. For the particles used, light ions, the position of interaction, called centroid of the cluster, in the detector is determined with sub-pixel spatial resolution at the few μm level. The resulting spatial image is produced as the map of the centroid of interaction of the single particles which can be displayed with spatial bin of size above this level. Given the low event rate by the microbeam flux, and limited detector live time, the spatial bin size in the image shown in Figure 3 was set to size 10 μm. The entire spatial cluster, which involves many of the detector pixels and extends to several hundreds of μm (see Figure 2), is not displayed in the imaging map of the microbeam scanning spot region.

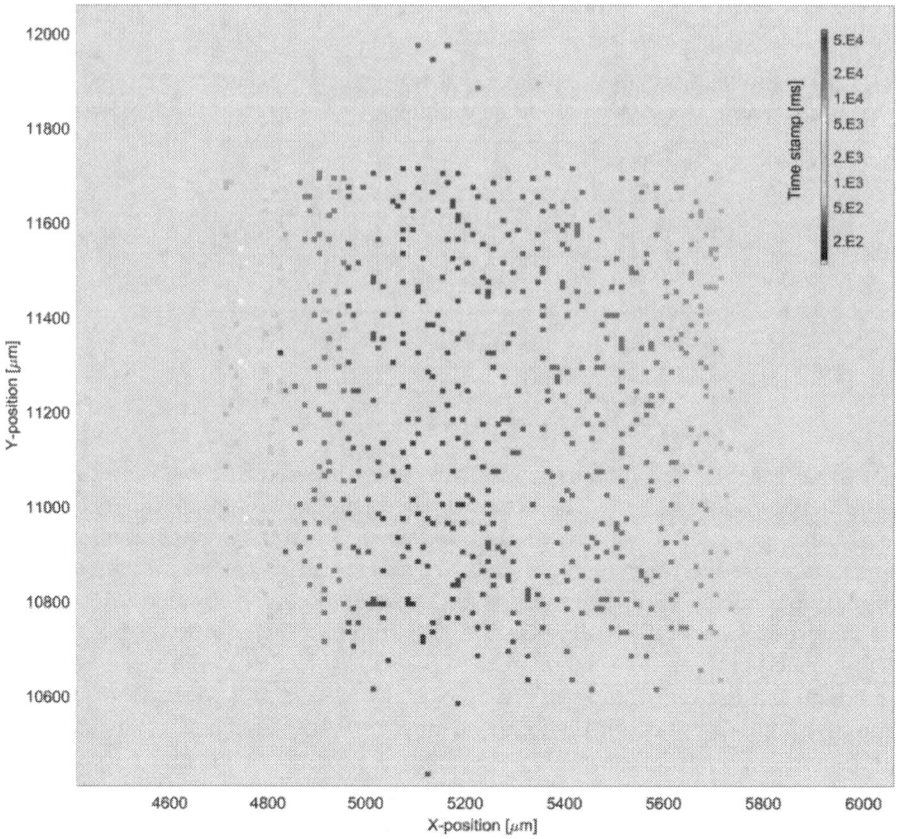

Fig. 3. Time-sensitive spatial distribution of single 10 MeV ^{12}C ions in a scanning microbeam of area 1 mm × 1 mm, scanning speed 4 μm/ms. Data shown for one full scan over the scanning spot area irradiated in 327 s (of which 113 s was the total live time of the detector). The time-of-detection of the single ions in the detector is shown in color in ms. Only the particle hit centroid position is displayed in the plot spatial binning size of 10 μm. Data registered by the MiniPIX Timepix detector operated in per-pixel time-of-arrival mode. Only a part of the detector pixel matrix area (1.4 %) is shown (30 × 30 pixels = 1.65 mm × 1.65 mm = 2.72 mm^2).

3.3 Spectral-correlated time histogram of delivered ion position

The detailed time distribution of the single particles registered by the detector can be evaluated. The time histograms of the position of the detected ions along the X and Y axis of the detector are shown in Figure 4. The same data of Figure 3 is used. The time range is partly extended (from 65 s to 80 s) which contains one entire spatial scanning over the microbeam spot area with additional 15 s of the next beam scan in order to show the spatial scanning pattern along the X and Y axis directions.

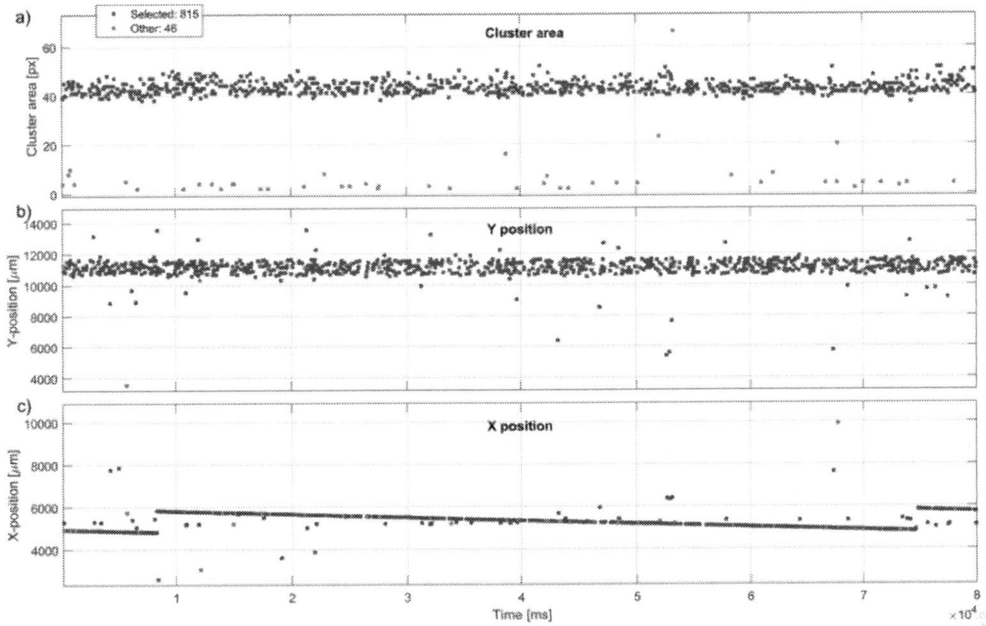

Fig. 4. Time histograms of detection of single ions by Timepix. The plots show the values of (a) cluster size, (b) Y position and (c) X position for the single particles. Results shown for the same data of Figure 3. Two groups of particles are resolved: primary beam particles – 10 MeV ^{12}C ions – blue (815) and other particles – red (46). The evaluated time interval, plotted on the horizontal axis, is extended to 80 s (from 65 s – interval of one complete spatial spot scan by the beam). See text.

Figure 4 includes the cluster size (number of pixels) of the registered particles in the detector. This quantity is proportional to the deposited energy. ^{12}C ions of 10 MeV and 4 MeV produce clusters of different area: of around 50 and 30 pixels, respectively – see Figure 2. Together with additional cluster morphology parameters such as cluster linearity and roundness, these quantities can be used to resolve particle-event types and thus characterize the composition of the radiation field [5]. Primary beam ions can be discriminated from background and unwanted events such as electrons, X rays, scattered particles and ions of different energy.

The correlated spatial- and time-information can be used to examine and characterize the deposited beam in terms of the composition (particle types) of the delivered beam, time and position of delivery with μs and μm level resolution, respectively.

3.4 Time-sensitive distance travelled of ion microbeam spot

The spatial trajectory of the microbeam spot delivery can be inspected and evaluated by Timepix at the μm spatial and μs time resolution level. We evaluated the beam *travelled distance*, expressed as the consecutive distance between delivered single ions on the detector sensor plane. The value of this quantity along the irradiated time is shown in Figure 5. Results are presented for the same data of Figure 3 (and of Figure 4 in the 65 s interval of an entire scan over the spot region). The travelled distance can be expressed in 2D (shown in black) or along the X- and Y- axis components (shown in blue and red, respectively).

The results shown, in Figure 5, Figure 4 and Figure 3, are produced by the particles detected in the detector. Thus, due to the detector readout dead time, they represent about one third of all irradiated particles by the beam during the same time interval evaluated.

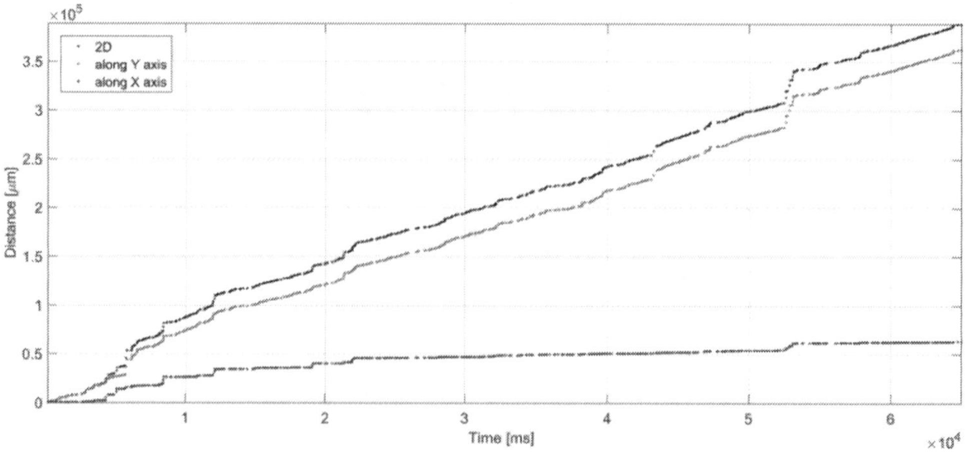

Fig. 5. Traveled distance of the 10 MeV ^{12}C microbeam for the data in Figure 3. The quantity (data points in black) can be also evaluated along the detector axis Y direction (red) and X direction (blue).

3.5 Ion microbeam imaging: particle rate map

We investigated the delivery and spatial visualization of a 4 MeV ^{12}C ion scanning microbeam of customized spatial pattern of scanning area 600 µm × 700 µm and scanning step of 3.9 µm (resulting from the beam spot binning 1 mm / 256 steps) – see Figure 6. The customized beam pattern was in the form of a pre-defined image produced by the Tandetron scanning microbeam system which was defined over a bit map such as those used in ion beam lithography.

Figure 6 shows the integrated particle count per image spatial bin registered by the detector. As in Figure 3, the image spatial bin size was also set to 10 µm. The data shown were collected in 5 min during which the beam performed several scans over the scanned area. A total of 3012 ions were registered by the detector during this interval. Over 10 k Timepix frames of acquisition time 10 ms were collected. The total detector frame exposure time (103 s) gives a total live time of 35%. Thus, the actual number of ions delivered is about twice as many as those registered. The number of non-empty spatial image bins, i.e., containing one or more events, in the entire imaged region displayed is 1827.

In this measurement the Timepix detector was operated in ToT per-pixel energy mode and readout in frame mode with acquisition time 10 ms. The detector frame acquisition time was set at this level in order to register single particles and avoid event pileup while maximizing detector exposure live time.

In a given Timepix frame, with the detector per-pixel operation set in energy mode, particles arriving at different detector regions are separated and can be distinguished. However, when the particles arrive to the same detector position in the given frame *event pileup* in the detector can occur. This means two or more particles appear registered as one particle of higher deposited energy in the detector giving rise to *event energy multiplets*. This effect in the detector can account for high-energy groups in the deposited energy spectrum (see Sec. 3.6).

The number of delivered ions (Figure 6) gives the event rate and total fluence in the evaluated region and scanned time interval. The beam spatial deposition and homogeneity are visualized and can be evaluated. It can be observed a distribution of values in the pattern image ranging from 0 (no events) up to six events per spatial bin. The beam composition in terms of main particle-type groups can be derived by the detector resolving power and pattern recognition analysis of the pixelated tracks [7]. In the results presented, all detected particles are processed and shown. The radiation field at the detector position was quite pure, containing primarily the microbeam particles, i.e., ^{12}C ions. If necessary, background and unwanted particles, such as electrons and X rays, can be filtered out by the detector particle-type resolving power and spectral-tracking sensitivity [7].

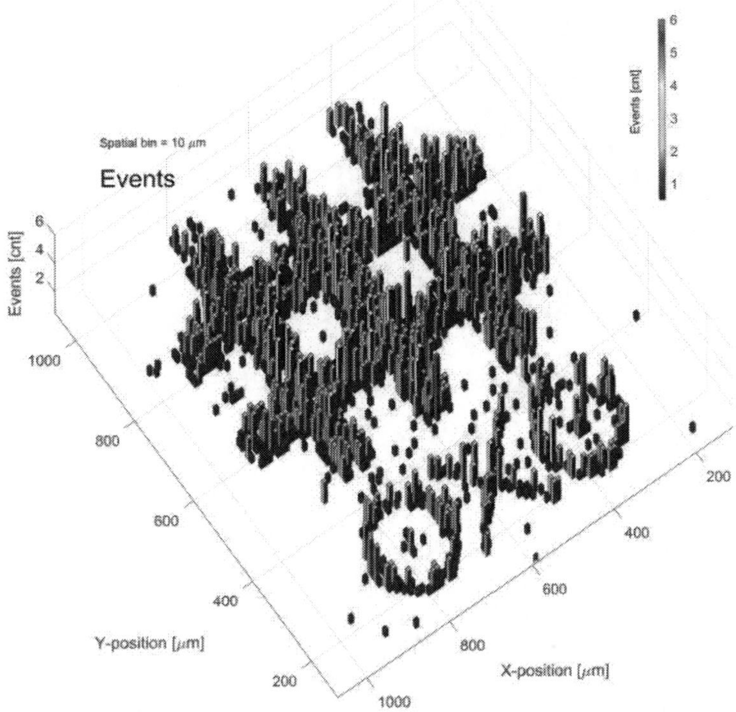

Fig. 6. Spatial distribution of ion delivery of a 4 MeV ^{12}C scanning microbeam with customized spatial pattern of size 600 μm × 700 μm registered by Timepix. Total beam irradiation time 5 min, during which time the beam scanned the spot region several times. The number of events (**event rate** – evaluated per image spatial bin, set of size 10 μm) is displayed in the vertical axis and in the color bar scale. A small portion of the pixel detector area (0.74%) is shown (22 × 22 detector pixels = 1.21 mm × 1.21 mm = 1.46 mm^2).

3.6 Ion microbeam imaging: deposited energy map

The spatial distribution of deposited energy by single particles can be evaluated also at the μm scale. The same data set used in Figure 6 is processed and displayed in terms of deposited energy according to the same spatial image binning as given in Figure 7. When two and more events were registered in different detector frames and correspond to the same spatial bin, the average value is displayed in the spatial image. Similar to the event rate image above, a range of values is observed over the scanned spot region. In terms of deposited energy several group values are registered by the detector in the range from 2 MeV to nearly 14 MeV. Most events in the detector were registered with the lowest energy group (see discussion below). In the evaluated data set, containing several scanning sweeps by the beam, most spatial bins in the pattern correspond to the lowest energy peak. The other higher energy multiplet groups are observed in decreasing incidence. There are also bins without any values which correspond to image spatial bins without any events – see discussion in Sec. 3.5.

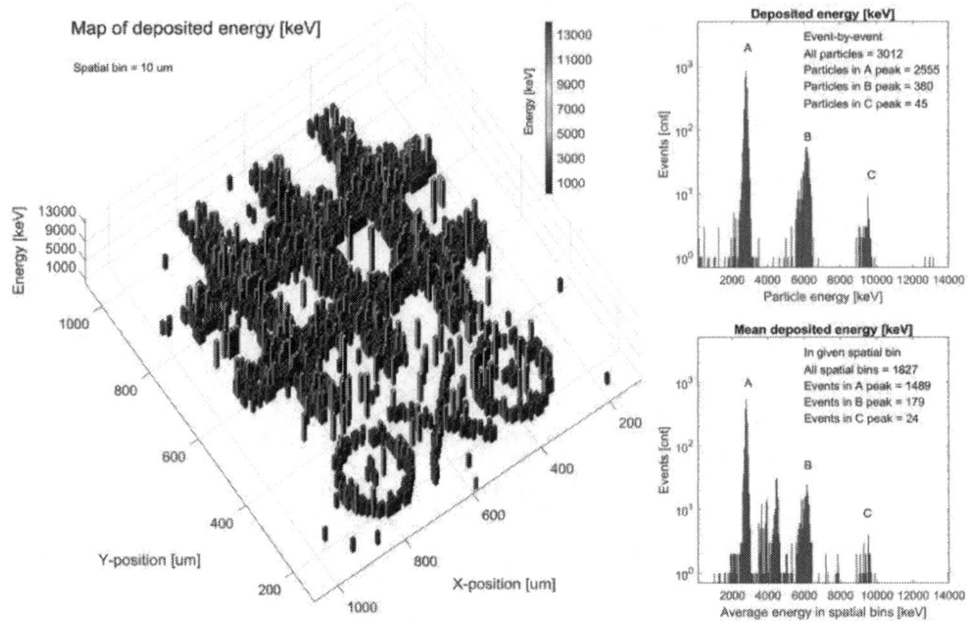

Fig. 7. Same as Figure 6 showing the ion **deposited energy** per spatial bin (displayed in the vertical axis and in color). For spatial bins registering two and more ions, the average value of ion deposited energy is displayed. The deposited energy spectrum of single ions is included (top right). The resulting distribution of averaged energies evaluated according to the spatial bins of the image is also shown (bottom right). See text.

3.7 Ion microbeam spectrometry: deposited energy spectrum

In addition to the spatial distribution of single particle delivery, the deposited energy spectrum can be measured and evaluated. The resulting distribution for the event-by-event processing is given in Figure 7 (top right). The spectrum measured by the detector shows a dominant low-energy component with peak centroid at around 2.8 MeV. This component corresponds to the primary beam mono-energetic particles. The difference in energy from the accelerator value (4 MeV ^{12}C ions) can result from non-negligible energy loss in the detector thin insensitive electrode layer ($\approx$ 1 μm, Aluminum) significant for heavy charged particles and the limited energy resolution of the detector especially in the high energy region. The detector per-pixel calibration is performed with X rays and gamma rays in the low energy region [12]. The detector energy resolution, for high-energy loss particles such as ions, is at the level of 8% [5,14]. Another source of loss of charge collected is the incomplete sensor depleted volume (estimated at around 60%), due to the insufficient applied bias. The low bias was chosen in order to suppress the per-pixel high-energy distortion and enhance the charge sharing effect.

In the event-by-event spectrum Timepix registers additional higher-energy groups. Their positions appear at multiple values of the first low-energy group. This points out to occurrence in the detector of multiplet events making three groups: doublets, triplet and a four-fold component of energies at around 6, 9.7 and 13 MeV, respectively. These higher-order groups registered by the detector appear as single particles of multiple-fold energy. They can however be two and more single particles arriving to the pixel detector within the detector acquisition time (10 ms) and in the same spatial position at the detector pixel size level, i.e., 55 μm. Such events cannot be resolved by Timepix operated in per-pixel energy mode. They could be resolved by operating the detector in counting mode, for which however the energy sensitivity of Timepix becomes limited.

Taking into account the imaging information, the spatially-correlated spectrum of deposited energy can be evaluated. In the case when more events are registered in the same image spatial bin (in the results shown the spatial bin is set to 10 μm), the averaged deposited energy per spatial bin is shown as the spectral image – Figure 7 (left) and as the energy spectrum – see Figure 7 (bottom right).

Due to the detector limited exposure live time, many of the delivered particles (about twice as many) are not registered. This effect in the detector can partly account for the inhomogeneity of ion registration over the scanned spatial pattern. Similarly, the total number of non-empty spatial bins for the data and interval processed (1827) is therefore smaller than the total number of particles registered by the detector (3012). The event occupancy values according to the large component groups, revealed in the event-by-event spectrum (Figure 7 top right), are included in the plot – see text inset in Figure 7. It can be noted an additional component appears between the first and second component of the event-by-event results. This results from the averaging of two and more events in a given spatial bin, which were registered with different energy values in the event-by-event data.

The deposited energy distributions results, processed as event-by-event particles or as spatially binned events (Figure 7) can be used to derive information on event rate, particle energy loss and deposited dose in the scanned spot region. Application of this technique is used in energy-sensitive particle radiography [13] in which few particles, even just one, are needed in each image bin to produce imaging contrast of thin samples. The correlated time information is also registered and can be evaluated for homogeneity of beam spot delivery.

The combined results provide a detailed and complete mapping of the delivered beam. Spatial and time distribution are measured in terms of number of particles, deposited energy, energy loss and time. For heavy charged particles the images can be produced with sub-pixel spatial resolution down to μm scale [12]. For heavy ions even sub-μm level spatial resolution can be achieved [14]. Spectral information of deposited energy is provided for each particle with energy resolution for charged particles at the 4 % level [7].

4 Conclusions

The spectral and time sensitive imaging of ion microbeams has been examined with the Timepix detector. Quantum imaging sensitivity is provided by the detector high granularity and dark current free detection at the pixel level. Beam spot spatial and time patterns at the μm and μs levels can be inspected and independently monitored with a miniaturized detector operating in vacuum and at room temperature. The position of single particles is determined with sub-pixel spatial resolution down to the μm level. The resulting image size and image spatial bin are given by the detector spatial resolution and event rate (total fluence). A complete and precise characterization of the delivered microbeam is provided in terms of beam/radiation field composition, particle flux, deposited energy and beam spot homogeneity in terms of spatial and time distribution. Ion microbeams of varying configuration can be inspected. The method can be applied for ion beam lithography and online beam profile monitoring of delivered particle beams of low intensity (flux below 10^5 particles/cm^2/s). At this particle flux no radiation damage or distortion of the detector spectral response is observed. For particle beams of higher energy, when the delivered particles have large range and cross the detector sensor thus producing elongated tracks, also the directional mapping of particles incident on the detector can be measured and evaluated. For the Timepix detector used, with a 300 μm thick silicon sensor, this case occurs for protons of energy 8 MeV and light ions of energy above 10 MeV/u.

In addition to the inspection of beam size and spatial- and time- profile, the divergence of particle beams can be also examined and imaged together with information on the beam composition and deposited dose distributions. Future work includes processing and evaluation of additional data with Tandetron microbeams at other settings and beam spot ranges. Also, the use of the newly available Timepix3 ASIC chip, which provides two per-pixel signals, enables to register simultaneously both the deposited energy and time at the pixel level. Timepix3 will enable to resolve pile-up and spatially-overlapping events in the detector. Also, more particles can be registered thanks to the Timepix3 hit-pixel readout stream architecture which greatly reduces detector readout dead time.

Work by C.G. an C.O. was performed in frame of Contract No. 40001250020/18/NL/GLC/hh by the European Space Agency. Measurements at the Tandetron accelerator were carried out in frame of the CANAM infrastructure of the NPI CAS Rez, Research Infrastructure project No. LM2015056 of the Ministry of Education, Youth and Sports of the Czech Republic. We thank Jack Miller from the Lawrence Berkeley National Laboratory, USA, for English proof-reading and comments. We are grateful to the reviewers for remarks and valuable suggestions.

References

1. O. Romanenko et al., Rev. Sci. Inst. **90**, 013701 (2019)
2. R. Ortega, et. al., NIM-B **181**, 475-479 (2001)
3. J. Baglin, App. Surface Science **258**, 4103-4111 (2012)
4. X. Llopart, R. Ballabriga, M. Campbell, NIM-A **581,** 485–494 (2007)
5. C. Granja, J. Jakubek, S. Polansky, et al., NIM-A **908**, 60-71 (2018)
6. C. Granja, K. Kudela, J. Jakubek, et al., NIM-A **911**, 142-152 (2018)
7. C. Granja, C. Oancea, J. Jakubek, et. al., NIM-A **988**, 164901 (2021)
8. C. Granja, J. Jakubek, M. Martisikova, et. al., JINST **13**, C11003 (2018)
9. V. Patera, A. Sarti, IEEE TNS **40**, 133-146 (2020)
10. T.Y. Hsiao, H. Niu, T.Y. Chen, C.H. Chen, MethodsX **7**, 100773 (2020)
11. A. Natochii, F. Murtas, W. Scandale, J. Alozy, JINST **14** P03018 (2019)
12. J. Jakubek, A. Cejnarova, T. Holy, et. al., NIM-A **591**, 155-158 (2008)
13. V. Olsansky, C. Granja, C. Oancea, et. al., these proceedings (2022) *submitted*
14. C. Granja, J. Jakubek, S. Pospisil, et. al., NIM-A **574**, 472-478 (2007)

EPJ Web of Conferences **261**, 02001 (2022) https://doi.org/10.1051/epjconf/202226102001
ANPC 2021

One-step 3D microstructuring of PMMA using MeV light ions

Oleksandr Romanenko[1*], *Adela* Jagerova[1,2], *Andrei* Borodkin[3], *Vladimir* Havranek[1] and *Anna* Mackova[1,2]

[1]Nuclear Physics Institute, Czech Academy of Sciences, 25068 Rez, Czech Republic
[2]J. E. Purkinje University, 40096 Usti nad Labem, Czech Republic
[3]Institute of Photonics and Electronics, Czech Academy of Sciences, 18251 Prague, Czech Republic

Abstract. The conventional procedure for creating 3D microstructures in resists by ion beam lithography consists of two stages – exposure and developing. However, single stage of manufacturing 3D structures in resist is also possible. Irradiation of PMMA can cause it to shrink. This feature of the polymer can be used for one-step three-dimensional microstructuring, which simplifies the manufacturing process. The shrinkage of PMMA film on a substrate has been extensively studied, while research on free-standing film is not comprehensive. The use of free-standing PMMA film allows the creation of a flexible material with 3D microstructures, which can be used in medicine, optics, and electronics. The question here is whether the results obtained for the PMMA film on the substrate are applicable to the free-standing film. Since the nature of shrinking is outgassing of volatile products, the film on the substrate has only one surface for the release of gases, while in the free-standing film, gases can be released from the sample from both sides. Therefore, the shrinking in the free-standing film occurs on both sides. The aim of this work is to study the shrinkage of the free-standing film and compare it with that of the film of the same thickness coated on the substrate.

Conventional ion beam lithography as well as direct (maskless) ion beam lithography follow two steps: 1) irradiation of material; 2) developing the material with chemical reagents. As result, 3D structures in the material can be created. PMMA, PDMS and SU-8 are the most used the polymers in ion beam lithography. It has been shown that microlenses can be formed in PDMS with direct ion beam lithography using heavy ions such as Nitrogen without chemical development [1]. This is possible to do due to the property of PDMS to shrink under irradiation. PMMA also shrinks when exposed to an ion beam [2,3]. Such behaviour of the polymers is explained by the leaving of volatile products from the material [4]. H_2, CO, CO_2 and CH_4 are the main gases produced in PMMA during irradiation, which subsequently diffuse from the material [5-7]. This feature can be exploited to simplify the production of the optical devices made of the polymer.

The flexibility of the optical devices is gaining in importance. This for example allows creating the digital cameras that mimic compound eyes commonly found in insects or spiders

* Corresponding author: romanenko@ujf.cas.cz

and are characterized by extremely large field of view angle, low aberrations, acute sensitivity to motion, and infinite depth of field [8]. In case of the polymer, a flexible sample can be obtained in two ways: i) using a flexible substrate; ii) making the sample thin enough to be flexible, which means producing a free-standing polymer film. If two polymer films of the same thickness are prepared by these two methods and exposed to ion beam irradiation, they will most likely have different shrinkage. This can be explained as follows: as already mentioned, the cause of shrinkage is the degassing of volatile products. These products can diffuse out of the sample from only one side in the case of the film on the substrate, while in the case of the free-standing film, they can diffuse from both sides of the sample, causing the polymer to shrink on both sides. This is schematically illustrated in Fig 1.

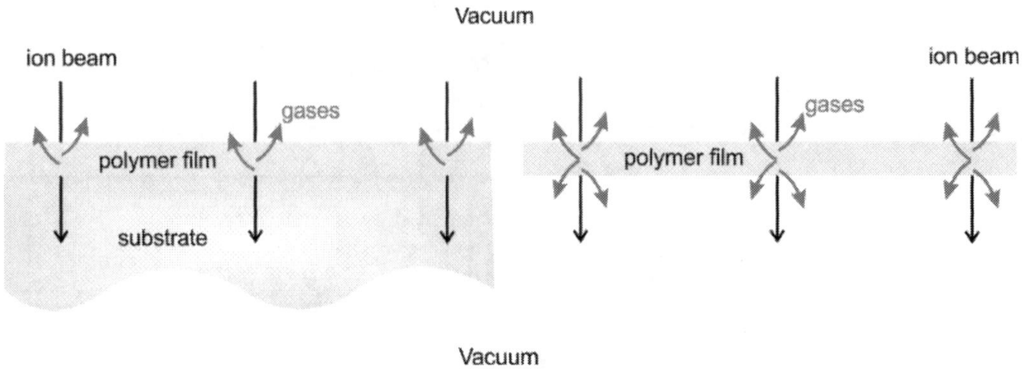

Fig. 1. Schematic illustration of the degassing of the polymer film.

In our work, we demonstrate the applicability of one-step three-dimensional microstructuring for the fabrication of the optical device created by irradiation alone. We also demonstrate the difference in structures created in the film on the substrate and the free-standing film irradiated under the same conditions. For our goals, we used PMMA because of its good optical transparency in visible light, easy handling and low cost.

Our samples were prepared as follows: 1 g of PMMA powder and 7.67 g of acetone were mixed in a vial with a magnetic stirrer. When PMMA was completely dissolved in acetone, the solution was spin-coated on the polyethylene and glass substrates using 400 rpm for 100 s. The PMMA film on polyethylene was subsequently peeled off to obtain the free-standing flexible PMMA film. The thickness of both the PMMA film on glass and the free-standing PMMA film was 20 μm. Irradiation was carried out on a nuclear microbeam facility at our institute[9]. 2 MeV proton beam focused into a spot of 0.77×2.37 μm^2 with a beam current of 68 pA were used to irradiate the samples. Diffraction grating of 100 lines/mm was chosen as the optical device to demonstrate the applicability of the method. The irradiation parameters were set as follows: the dose – 400 nC/mm^2 (2.5E14 protons/cm^2), scan size – 1024×1024 pixels, pixel dimension – 1×1 μm^2. To reduce the heating effect caused by irradiation, the irradiated pattern was created by scanning in 11 loops. This increases the time required for irradiation, but reduces the heat load well. The total time spent on one diffraction grating was 632 s.

The thickness of the film and the energy of the protons were chosen in such a way that the protons transfer their energy to PMMA uniformly over the entire depth of the film. SRIM program [10] for modelling the interaction of ions with the material shows that the difference of linear energy transfer (LET) of 2 MeV protons in PMMA is about 15 % between the front and the back side of the film (Fig. 2).

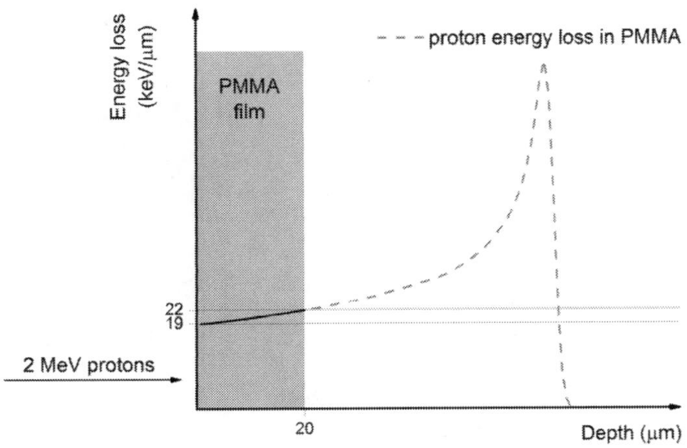

Fig. 2. Linear Energy Transfer of 2 MeV protons in PMMA. The solid line shows the protons energy losses in the prepared PMMA film.

Optical analysis of the irradiated samples revealed distinct lines corresponding to the places of interaction of the proton beam with the material (Fig. 3). The quality of the created gratings were good, although some inclusions were noticed. We assume that the appearance of inclusions was caused by fluctuation in the proton beam current. Despite the 11-loops scan, which smooths out fluctuations in the beam current and thus makes the irradiation load more uniform, some pixels received noticeably higher fluence than others. This can be explained in the following way: The time spent in every pixel was the same, therefore those pixels that were exposed to the beam when it had the high current accumulated the higher fluence compared to others. If this happened in every loop, the final fluence in such pixels was greater than in others. This problem can be eliminated by increasing the number of scan loops, but this will increase the irradiation time [2]. Visually, the width of the lines appears to be greater than the beam width of 0.77 μm, although precise lines width measurements were not possible on the obtained optical images. For this it was better to use AFM.

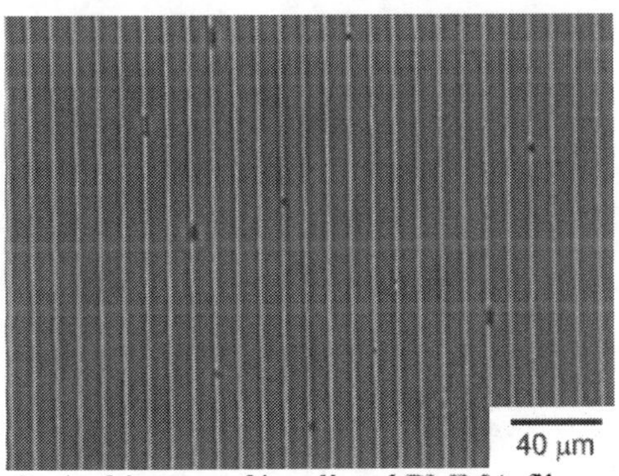

Fig. 3. Optical image of irradiated PMMA film on glass.

AFM analysis was used to investigate the topography of the irradiated area. As can be seen in Fig. 4, shrinkage in the film coated on glass was more pronounced than in the free-standing film. The profile of the grating created in the film on glass shows that the depth of the grooves was 160 nm (Fig. 5). In case of the free-standing film, the depths of the grooves were 9 and 11 nm on the front and back sides of the film, respectively. The difference in grooves depth between front and back sides is explained by LET, that increases with depth. The total shrinkage of PMMA in the free-standing film is 20 nm, which is almost an order of magnitude lower than that for PMMA film on glass. The reason of such big difference is unclear.

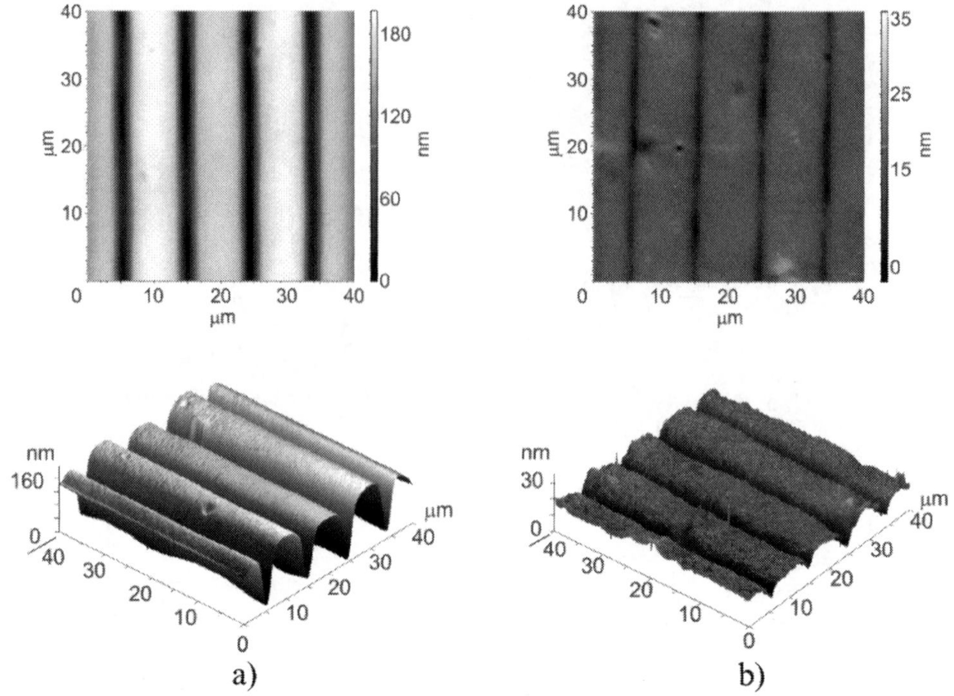

Fig. 4. AFM images of the PMMA films irradiated with the same condition: a) PMMA film on glass; b) free-standing PMMA film (front side).

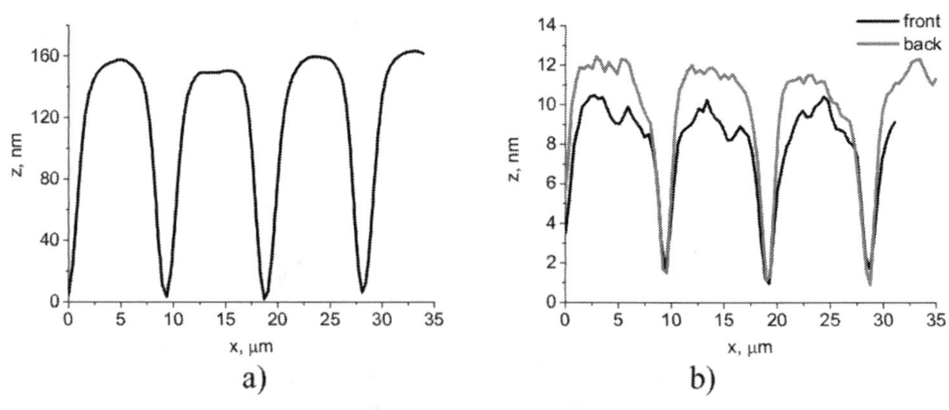

Fig. 5. Profiles of the created diffraction gratings showing PMMA shrinkage: a) PMMA film on glass; b) free-standing PMMA film.

From Fig. 4 it seems that the roughness of unirradiated area in the free-standing film is higher than that in the film on glass, but this is just illusion caused by the difference in scale. Analysis of unirradiated area revealed that its roughness is about 1.6 nm for both samples.

Gaussian fit, implemented in the OriginPro software, was used to analyze the profile of the grooves in PMMA. It was found that FWHM of the grooves in PMMA on glass is 2.23 µm, and in free-standing PMMA – 1.76 µm. This difference is explained by the different depth of the grooves. When the polymer shrinks in the place exposed to the ion beam, it pulls along the unirradiated part of the material adjacent to it. Thus, the more the polymer shrinks, the more it pulls the undamaged material with it, which leads to a widening of the groove.

To check the quality of the created diffraction grating, monochromatic light with vertical polarization and a wavelength of 520 nm was utilized. An interference pattern was obtained on a screen located at a distance of 50 cm from the diffraction grating (Fig. 6). Close look on the interference pattern shows parasitic flare. This was observed for both samples, although in the case of the PMMA film on glass, parasitic flare was less noticeable. Such parasitic flare is caused by defects in the diffraction grating. The quality of the diffraction grating can be assessed by the intensity of the fringes. Table 1 shows the measured fringe intensities. As can be seen, the ratio between the zeroth and other orders is greater in the PMMA film on glass, which means a better quality of diffraction grating compared to a free-standing PMMA film.

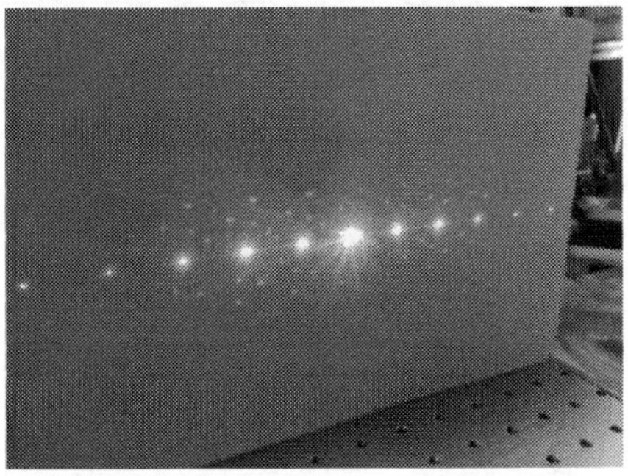

Fig. 6. Interference pattern obtained with created diffraction grating in the free-standing PMMA film.

Table 1. Intensity of fringes in the interference patterns obtained with the created diffraction gratings.

Fringe order	Intensity, µW Free-standing PMMA film	Intensity, µW PMMA film on glass
0	600	860
+1	112	22
-1	114	27
+2	55	27
-2	59	30
+3	12	12
-3	14	12

Interference pattern allows the period of diffraction grating be evaluated. Calculated period for every order is presented in Table 2. The fluctuation of the values is less than 5% and is associated with an error in the selection of the center of the fringe during the measurement. The pattern for irradiation was 100 lines/mm which gives the period of the diffraction grating of 10 µm. Thus, the created diffraction grating reproduces the original pattern with high accuracy.

To summarize, it was shown that one-step 3D microstructuring of the polymer using MeV light ions is viable. 3D microstructures in PMMA have been successfully created with 2 MeV protons in a single step. The application of this approach has been demonstrated by the example of the production of the optical device – the diffraction grating. The fabricated diffraction grating showed good agreement with the original pattern. The structures quality was good, with some minor artifacts, which can be eliminated by changing the scanning method. One-step microstructuring simplifies the manufacturing procedure and has great potential for prototyping new optical devices.

Table 2. The period of the diffraction gratings calculated from the interference pattern.

Fringe order	Free-standing PMMA film		PMMA film on glass	
	Distance from zeroth order, mm	Calculated grating period, µm	Distance from zeroth order, mm	Calculated grating period, µm
+1	26	10.01	25	10.41
-1	26	10.01	26	10.01
+2	53	9.88	52	10.07
-2	53	9.88	53	9.88
+3	81	9.79	79	10.03
-3	79	10.03	81	9.79

The research has been carried out at the CANAM (Centre of Accelerators and Nuclear Analytical Methods) infrastructure LM 2015056. This publication was supported by OP RDE, MEYS, Czech Republic under the project CANAM OP, CZ.02.1.01/0.0/0.0/16_013/0001812 and by the Czech Science Foundation (GACR No. 19-02482S)

References

1. G. U. L. Nagy, V. Lavrentiev, I. Banyasz, S. Z. Szilasi, V. Havranek, V. Vosecek, R. Huszank, I. Rajta, Thin Solid Films **636**, 634 (2017).
2. O. Romanenko, V. Havranek, P. Malinsky, P. Slepicka, J. Stammers, V. Svorcik, A. Mackova, D. Fajstavr, Nucl. Instrum. Methods B **461** 175 (2019)
3. T. C. Sum, A. A. Bettiol, H. L. Seng, I. Rajta, J. A.van Kan, F. Watt, Nucl. Instrum. Methods B **210** 266 (2003)
4. F. Schrempel, Y.-S. Kim, W. Witthuhn, Appl. Surf. Sci. **189** 102 (2002)
5. B. Pignataro, M.E. FragalB, O. Puglisi, Nucl. Instrum. Methods B **131** 141 (1997)
6. E. H. Lee, G. R. Rao, L. K. Mansur, Radiat Phys Chem **55** 293 (1999)
7. Z. Chang, J. A. LaVerne, Radiat. Phys. Chem. **62** 19 (2001)
8. Y. M. Song, Y. Xie, V. Malyarchuk, J. Xiao, I. Jung, K. Choi, Z. Liu, H. Park, C. Lu, R. Kim, R. Li, K. B. Crozier, Y. Huang, J. A. Rogers, Nature **497,** 95–99 (2013)
9. O. Romanenko, V. Havranek, A. Mackova, M. Davidkova, M. Cutroneo, A. G. Ponomarev, G. Nagy, J. Stammers, Rev. Scient. Instr. **90** 013701 (2019)
10. J. F. Ziegler, M. D. Ziegler, J. P. Biersack, Nucl. Instrum. Methods B **268** 1818 (2010)

Properties of polyamide 6 and polyvinylidene fluoride nanofibers irradiated by H^+ ions

Eva Štěpanovská[1,2*], *Petr* Malinský[1,2], *Jindřich* Matoušek[2], *David* Poustka[2], *Anna* Macková[1,2]

[1]Nuclear Physics Institute of the Czech Academy of Science, *v. v. i., 250 68 Rez, Czech Republic*

[2]Department of Physics, Faculty of Science, J.E. Purkyne University, Pasteurova 3632/15, 400 96 Ústí nad Labem, Czech Republic

Abstract. This work deals with the modification of polymeric nanofibers of polyamide 6 (PA6) and polyvinylidene fluoride (PVDF) which were formed by electrospinning process. After the manufacturing process, the polymer nanofibers were exposed to the implantation of 1 MeV H^+ ions on a tandem accelerator Tandetron MC 4130. The ion implantation was provided with different ion fluences (1.8; 3.7; 5.6)$\times 10^{14}$ cm^{-2}. Ion implantation of polymer nanofibers can modify their functional properties due to ion interaction with nanofibers changing their structure and elemental composition. H-ion interaction with nanofibers was simulated by SRIM program which shows the modification of polymers by prevailing electronic stopping. Rutherford Back-Scattering spectrometry (RBS) and Elastic Recoil Detection Analysis (ERDA) show distinct elemental modification in the irradiated layer of PVDF and PA6 nanofibers. The changes in surface chemistry was identified by X-ray Photoelectron Spectroscopy (XPS). The identified chemical changes contributed to the changes of electrical properties (increase of electrical conductivity) being measured by the standard two-point method.

1 Introduction

Polymer nanofibers are a widely studied material. Due to their large effective area and porous structure, electrospun nanofibers can be employed in a wide range of applications [1,2]. When the modification of nanofibers leads to increase of their electrical conductivity and morphological changes, they can be used in fields such as electronics [3] or sensorics [4]. Although nanofibers have a diameter in the range of 1-100 nm, they mostly retain mechanical properties of bulk polymer. However, most modifications (e.g plasma, UV-light etc.) affect only chemical bonds on the surface. In this work, the modification of nylon 6 (PA6) and polyvinylidene fluoride (PVDF) nanofibers by the ion beam irradiation is followed. Ion beam interaction with nanofibers can change their electronic structure, chemical composition and it can lead to a change in polymer crystallinity, nanofiber morphology and electrical properties [5]. We have chosen irradiation by 1 MeV H^+ ions into polymer nanofibers, where

* Corresponding author: stepanovska@ujf.cas.cz

mainly electronic stopping is employed in the modified layer withouth significant disruptive ballistic processes and experiencing high penetration depth.

2 Experimental

For our purposes, electrostatic spinning was performed on the NS 1WS500U device (Elmarco, CZ). PVDF fibres were formed from 12 wt. PVDF solution. This solution was performed by mixing of 12 g of PVDF powder (Solvay, Belgium), 86.8 g of N, N-dimethylformamide (Penta, CZ) and 1.2 g of tetrabuthylammonium bromide (Sigma Aldrich, UK). The mixture was heated and stirred at 70 °C for 5 hours. The procedure for the formation of PA6 solution is given e.g. in [6]. The electrostatic spinning parameters are given in Tab.1.

Table. 1. Parameters of electrospinning on the NS 1WS500U device.

Parameters of electrospinning	PVDF	PA6
Distance of electrodes (mm)	180	240
Cathode voltage (V)	10	30
Anode voltage (V)	50	60
Winding of spoundbound (mm/min)	45	30
Humidity (%)	27	38
Temperature (K)	294.85	297.65

The prepared nanotextiles of PA6 and PVDF were further modified by implantation of H^+ ions with 1 MeV energy at the implantation fluences (1,8: 3,7: 5,6) × 10^{14} cm^{-2}. The ion beam was steered over area of 5×5 cm, directed vertically on the sample surface. The ion current during the irradiation was kept ~15-45 $nA.cm^{-2}$.

The PA6 and PVDF elemental composition was studied on the 1 MeV H^+ ions-irradiated samples using RBS and ERDA. The RBS/ERDA spectra were collected using 1.7 MeV He^+ ion beams. An Ultra-Ortec PIPS detector recorded backscattered He ions in Cornell geometry with a laboratory scattering angle of 170°. The ERDA primary beam incoming angle was 75° and the scattering angle was 30°. The recoiled particles were recorded by Canberra PIPS detector covered by 12 μm Mylar foil. The ion currents used during the RBS and ERDA analyses were ~5 nA. Several RBS spectra were collected at different sample parts to reduce the effects of sample degradation, with the final spectrum obtained by summing the individual spectra. The elemental concentration was analysed using the SIMNRA code [7].

The ion projected ranges and energy losses of the H^+ 1 MeV ion beam in the PA6 and PVDF nanofibers were calculated by the SRIM full-cascade Monte-Carlo simulation [8], version Stopping power SRIM-2008 . The elemental composition of the PA6 and PVDF pristine sample used for the SRIM simulation was determined by RBS and ERDA (PA6: C – 58 %, H – 22%, O – 9 %, N – 11 % and PVDF: C – 62 %, H – 12 %, O – 3 %, F – 24 %). The density of PA6 (1.084 g/cm^3) was used from publication [9] and the density of PVDF (1.78 g/cm^3) was used from publication [10].

3 Results and discussion

3.1 SRIM

The SRIM-calculated electronic (S_e) and nuclear (S_n) stopping in the near-surface layer, the energy stopping ratio (S_e/S_n) and the projected range (R_P) of H^+ ions are presented in Table 2. Furthermore, the integral of electronic ($S_e(\int)$) and nuclear ($S_n(\int)$) stopping was determined in order to find the total electronic and nuclear energy loss of H-ions on the whole penetration range.

Table. 2. Data obtained from SRIM: R_P is the projected range of H^+ 1MeV ions, ΔR_P is the standard deviation of the ion depth distribution , S_e is the value of electronic stopping at the surface for a given polymer and S_n is the value of nuclear stopping, S_e/S_n is the ratio between electronic and nuclear stopping at the surface, $S_e(\int)$ is integral value of electronic stopping in the whole penetrated volume, $S_n(\int)$ is the value of nuclear stopping in the whole penetrated volume and $S_e(\int)/S_n(\int)$ is the ratio between electronic and nuclear integral stopping.

Nanofibers/ Ions	R_P (μm)	ΔR_P (μm)	S_e (eV/nm)	S_n (eV/nm)	S_e/S_n (eV/nm)	$S_e(\int)$ (eV/nm)	$S_n(\int)$ (eV/nm)	$S_e(\int)$/ $S_n(\int)$ (eV/nm)
PA6/H^+	26.8	6.7	24.26	0.01	2426	99,4×10^4	1044.7	952
PVDF/H^+	17.1	4.2	38.44	0.01	3844	99,2×10^4	1071.7	926

The depth profiles of S_e and S_n ion energy losses together with 1 MeV H^+ depth distribution in PA6 and PVDF are depicted in Fig. 1a and 1b, respectively. The 1 MeV H^+ ions mostly lose their energy through the electronic stopping processes in both nanofiber layers (see Table 2 and Fig. 1) and the ratios of the electronic (S_e) and nuclear (S_n) stopping power of 1 MeV H^+ ions in PA6 and PVDF show the prevailing electronic stopping as well. During this process, electrons are released and chemical bonds are broken, the ejections of matrix atoms is rather unlike due to the low mass of H^+ ions undergoing negligible nuclear stopping. H-ion projected ranges don't cover the whole thickness of the nanofiber layers being in total about 66 μm for PA6 and about 60 μm for PVDF.

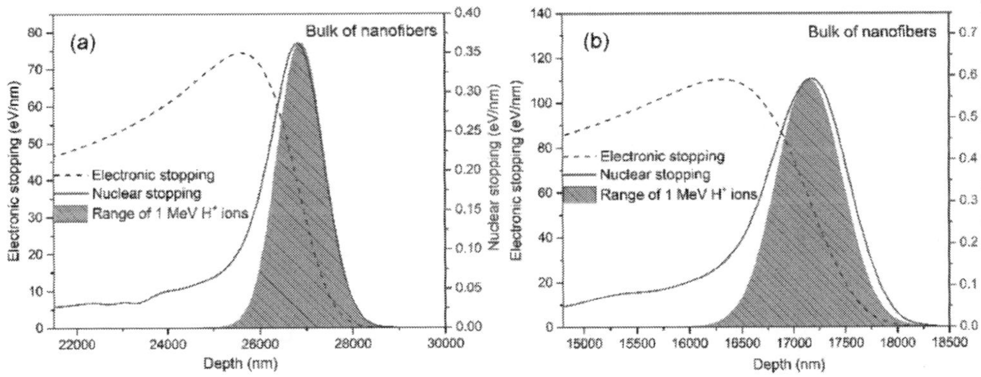

Fig. 1. The SRIM calculation of the electronic and nuclear energy stopping of 1 MeV H^+ ions vs. depth and the 1 MeV H^+ ion depth profiles in PA6 (a) and PVDF (b).

3.2 RBS and ERDA analysis of elemental compositions

The elemental composition of PA6 and PVDF nanofibers before and after the H-ion irradiation was determined using RBS and ERDA analytical methods. The accessible information depth of RBS using 1.7 MeV He-ions in both polymers is about 1 μm. The ERDA provides information up to the maximum depth of about 0.5 μm. RBS and ERDA spectra for the pristine and irradiated samples with the H-ion fluence of 3.75×10^{14} cm^{-2} are shown in Fig. 2(a), (b) for PA6 and (c), (d) for PVDF, respectively. The compositional changes are observed in PA6 after the H-ion irradiation (see Fig. 2(a) and (b)) mainly for C and H content. The composition of the irradiated nanofiber layers for all H-ion implantation fluences are summarized in Table 3. The C/O and C/N ratio slightly increases with increasing ion fluence in PA6, carbon content enhancement occurs already for the lowest ion implantation fluence and then for the higher irradiation fluences the carbon concentration stays constant. Hydrogen content is significantly decreased with the increasing H-ion fluence in the whole investigated surface nanofiber layer about 0.5 μm. PVDF (see Fig. 2(c) and (d)) exhibits a more significant change in elemental composition after the H-ion irradiation.

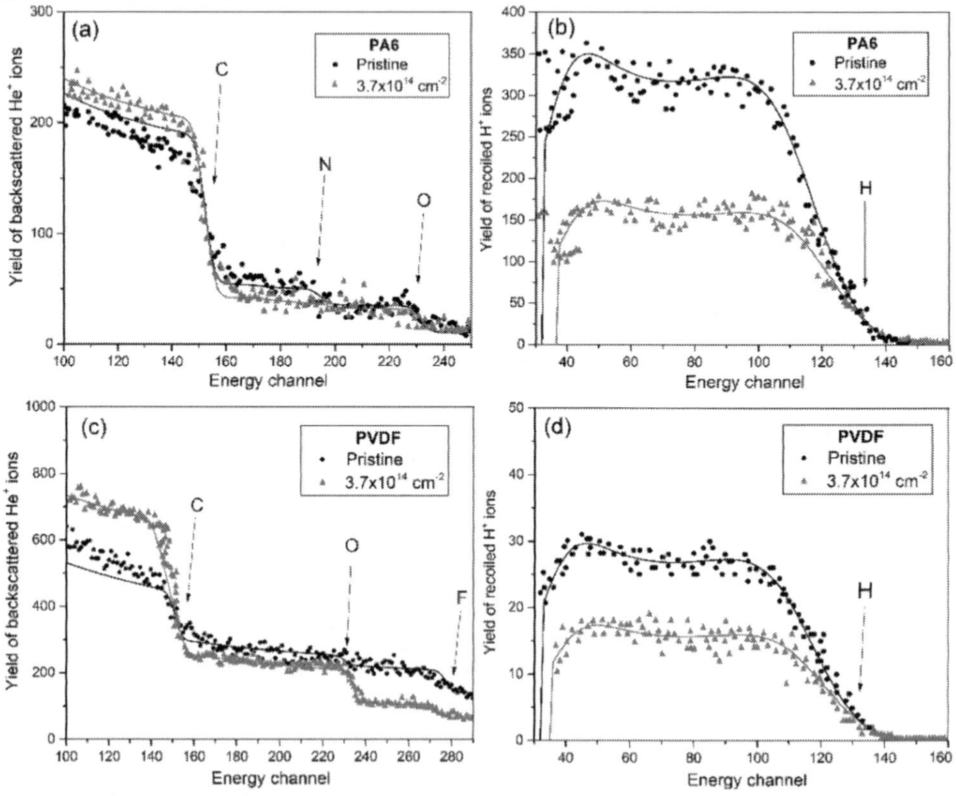

Fig. 2. RBS experimental (points) and SIMNRA simulated (line) spectra for PA6 (a) and PVDF (c) are presented together with ERDA experimental (SIMNRA simulated (line) spectra for PA6 (b) and PVDF (d)).

It should be noted that the oxygen found in the PVDF nanofibers is present due to the blending of the PVDF polymer with DMF (see chapter 2), because oxygen is not present in the PVDF stoichiometry. In PVDF, C/F ratio increases, and significant carbon concentration enhancement occurs with the increased H-ion implantation fluence simultaneously with H, F depletion (see Table 4). For avoiding unintentional effect of the He-ion irradiation during the RBS and ERDA analysis the typical He-ion current about ~ 5 nA was used and to reduce effects of the sample degradation, several spectra were measured on different beam spots and the final spectrum was obtained by summing the individual spectra.

Table 3. The PA6 stoichiometry before and after the H-ion irradiation determined using RBS and ERDA methods.

	Atomic concentration (%)						
Substrate	C	O	H	N	Impurities (Na, P, K, Mg)	C/O	C/N
PA6 Pristine	58 ± 5	9 ± 2	22 ± 4	11 ± 1	0.1	6	6
PA6/H^+ 1.8x10^{14} cm^{-2}	69 ± 6	8 ± 2	15 ± 3	8 ± 1	0.6	9	9
PA6/H^+ 3.7x10^{14} cm^{-2}	69 ± 6	8 ± 2	14 ± 3	7 ± 1	2	9	10
PA6/H^+ 5.6x10^{14} cm^{-2}	69 ± 6	7 ± 2	15 ± 3	6 ± 1	3	10	12

Table 4. The PVDF stoichiometry before and after the H-ion irradiation determined using RBS and ERDA methods.

Atomic concentration (%)						
Substrate	C	O	H	F	Impurities (Na, P, K, Mg)	C/F
PVDF Pristine	62 ± 5	3	12 ± 2	24 ± 3	0.1	3
PVDF/H^+ 1.8x10^{14} cm^{-2}	87 ± 8	1	5 ± 1	7 ± 1	0.9	12
PVDF/H^+ 3.7x10^{14} cm^{-2}	86 ± 8	1	4 ± 1	8 ± 1	1.0	10
PVDF/H^+ 5.6x10^{14} cm^{-2}	88 ± 8	1	4 ± 1	7 ± 1	0.6	13

3.3 X-ray photoelectron spectroscopy

X-ray photoelectron spectroscopy (XPS), providing information from only a few-nanometre-thick surface layer, was used to examine surface chemistry modification and to identify functional groups on PA6 and PVDF surfaces before and after H-ion irradiation. Fig. 3 shows the C 1s peak and its deconvolution with identification of individual contribution of chemical bonds, where Fig. 3(a) shows the spectrum of C 1s peak of unmodified PA6 nanofibers, Fig. 3(b) PA6 nanofibers irradiated with H-ions, ion implantation fluence of 1.8×10^{14} cm^{-2} and Fig. 3(c) PA6 nanofibers irradiated with H-ions with ion implantation fluence of 5.6×10^{14} cm^{-2}. The C 1s peak of pristine PA6 nanofibers has three different carbon bonding states (Fig. 3): C-C (284.5 eV), C-N, C-O (286.3 eV) and C=O (287.7 eV). These chemical bonds remained present in all PA6 nanofibers after the H-ion implantation. The binding energy values with the percentage of appropriate chemical bonds are presented in Tab. 5. The percentage of C-C bonds was reduced at the lowest ion fluence, while C-N and C-O bonds were increased. With increasing ion fluence, an increase in C-C bonds and a decrease in C-O and C-N bonds were observed. The C=O bonds remains practically constant.

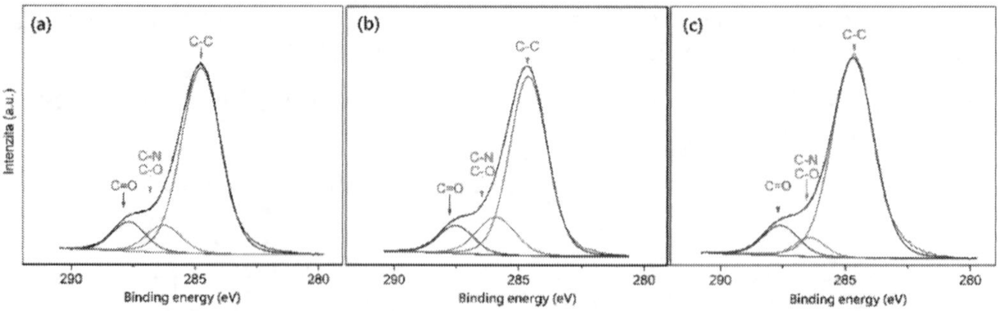

Fig. 3. The deconvolution of the C 1s peak from XPS spectra of (a) pristine PA6, (b) modified PA6 nanofibers with the ion implantation fluence 1.8×10^{14} cm^{-2} and (c) modified PA6 nanofibers with the with ion implantation fluence 5.6×10^{14} cm^{-2}.

Fig. 4 shows the C 1s peak and its deconvolution with identification of individual contribution of chemical bonds in PVDF, where Fig. 4(a) shows the spectrum of C 1s peak of the unmodified PVDF nanofibers, Fig. 4(b) the modified PVDF nanofibers with H^+ ion implantation fluence of 1.8×10^{14} cm^{-2} and Fig. 4(c) the modified nanofibers PVDF by H^+ ion implantation fluence of 5.6×10^{14} cm^{-2}. The C 1s peak of pristine PVDF nanofibers has four different carbon bonding states (Fig. 4a): CH_2, C-C (284.5 eV), C-F (289.6 eV), CF_2 (291.7 eV) and CF_3 (294.6 eV). These chemical bonds remained present in all PVDF nanofibers after H-ion implantation.

Table 5. Position and abundance of the individual components of the C 1s peak in pristine and H-ion irradiated PA6 nanofibers.

PA6/H^+			
Sample	**C-C**	**C-N, C-O**	**C=O**
	Binding energy (eV) / Percentage (%)		
Pristine	284.8/79.0	286.3/10.5	287.7/10.5
1.8×10^{14} cm^{-2} (H^+)	284.8/73.3	286.1/16.1	287.7/10.6
5.6×10^{14} cm^{-2} (H^+)	284.8/83.5	286.5/5.8	287.7/10.7

The binding energy values with the percentage of particular bonds in PVDF are presented in Tab. 6. Fig. 4a and 4b shows a decrease in the signal intensity of all fluorine-containing bonds after implantation of H-ions. In Tab. 6, an increase in C-C, CH_2 bonds was noted, while all fluorine-containing bonds decreased with the increasing H-ion implantation fluence. During irradiation, the enhancement of C-C bonds occurs on the surface of the polymer nanotextile at the expense of defluorination. The increasing carbonization with decrease of fluorine was also confirmed by the RBS analysis, where the trend with increased H-ion irradiation fluence is the same.

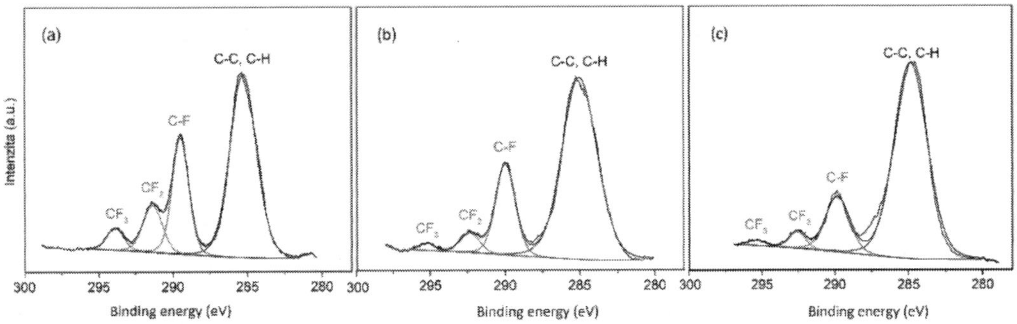

Fig. 4. The deconvolution of the C 1s peak from XPS spectra of pristine PVDF (a), modified PVDF nanofibers with H-ion implantation fluence 1.8×10^{14} cm^{-2} (b) and (c) modified PA6 nanofibers with H-ion implantation fluence 5.6×10^{14} cm^{-2}.

Table 6. Position and abundance of the individual components of the C 1s peak in pristine and H-ion irradiated PVDF nanofibers .

PVDF/H^+				
Sample	**C-C, CH_2**	**C-F**	**CF_2**	**CF_3**
	Binding energy (eV) / Percentage (%)			
Pristine	284.8/59.9	289.6/24.2	291.7/11.3	294.6/4.6
1.8×10^{14} cm^{-2} (H^+)	284.8/79.4	289.7/16.1	292.5/3.3	295.5/1.2
5.6×10^{14} cm^{-2} (H^+)	284.8/73.8	289.8/20.5	292.1/4.4	295.0/1.3

3.4 Electrical properties

Changes in the elemental composition and chemical bonds affected the electrical properties of the modified nanofibers. Fig. 5 shows the IV-characteristics and sheet resistance of PA6 (a, b) and PVDF (c, d) nanofibers irradiated with the different H-ion fluences. As illustrated in Fig. 5(a) and 5(b), with the exception of the H-ion irradiated PA6 nanofibers with a ion fluence of 3.7x10^{14} cm^{-2}, it can be said that the sheet resistance is a decreasing after H-ion irradiation and it can be concluded that the predominant electronic stopping of H-ions (see Fig. 1(a)) leads to the increase of C-C bonds and it improves the conductivity of the polymer nanotextile [11]. The values of measured sheet resistances are comparable to the values in literature e.g. [12]. Fig. 5(d) shows a plot of the sheet resistance measured on irradiated PVDF nanofibers. It is evident that the sheet restistance decreases after the H-ion irradiation at the lowest ion fluence. The sheet resistance increases again with the further increase of ion fluence, but it is still lower than the value of pristine PVDF nanofibers. The slight increase in sheet resistance, with increasing ion fluence, may be due to the deteriorated morphology of the nanofibers. It can be assumed that the H-ion irradiation enhances the conductivity of PVDF, but also disrupts the morphology of the nanofibers.

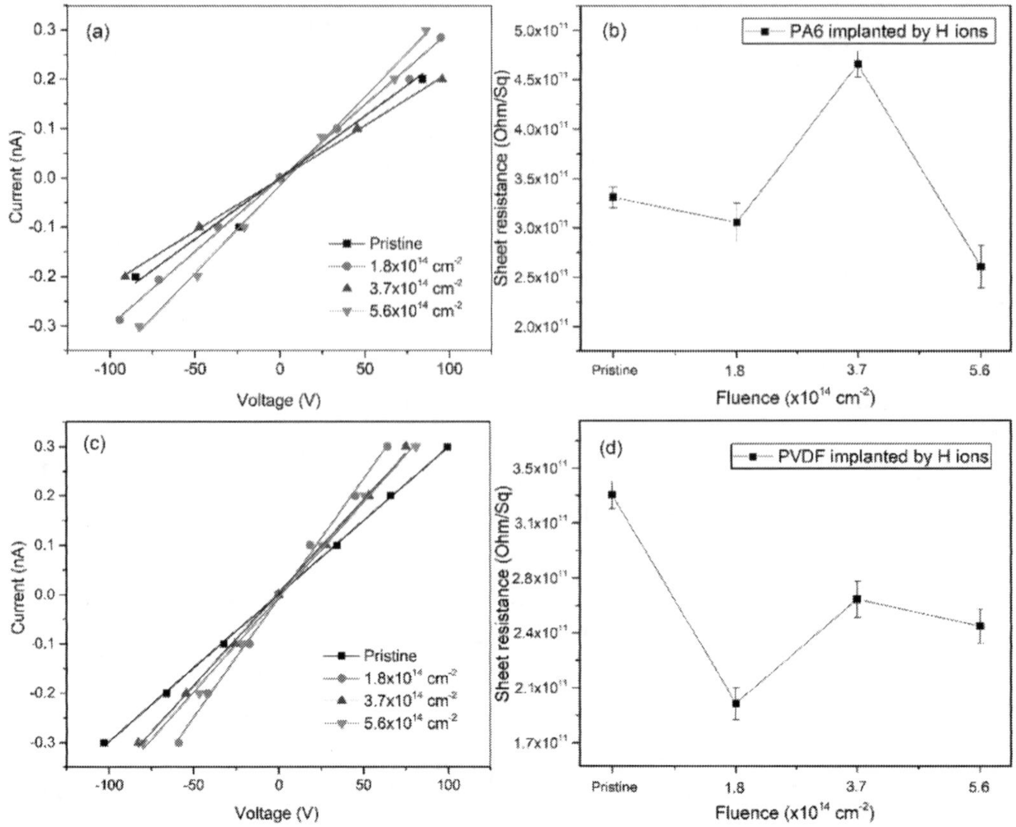

Fig. 5. IV-characteristics of modified PA6 (a) and PVDF nanofibers (c) with sheet resistance of the H-ion irradiated PA6 (b) and PVDF (d) using the various H-ion irradiation fluences.

4 Conclusions

In this work, the polymer nanofibers PA6 and PVDF irradiated by H^+ ions with different fluences (1.8; 3.7; 5.6)$\times 10^{14}$ cm^{-2} were studied. Using spectrometric and spectroscopic techniques (RBS, ERDA, XPS), changes in the elemental and chemical composition were observed which were caused mainly by electronic stopping of H-ions, as demonstrated by the SRIM calculations. Due to the predominant electronic stopping, the carbon content enhancement occurred in the PA6 polymer, accompanied by N and H depletion. The decrease of C-N bonds on the surface and the increase of C-C bonds was confirmed by XPS.

In the case of PVDF, there were more significant chemical and elemental changes than in PA6. RBS confirmed increase of carbon content with strong defluorination dependent on the increasing ion implantation fluence. The assumption of carbonization and defluorination was also confirmed by the XPS method.

In both PA6 and PVDF, there was a slight increase in electrical conductivity. In both cases, the conductivity was higher for the lower ion fluence and with increasing ion fluence, the conductivity decreased again. This conductivity reduction was probably caused by the morphological degradation of polymeric nanofibers. In general, PVDF exhibited more significant correlation of electrical conductivity increase with the increased H-ion fluence compared to PA6. This fact could be connected to more significant elemental modification and chemical bond disruption in PVDF compared to PA6, as it was shown by RBS and XPS.

Acknowledgement

The research has been carried out at the CANAM (Centre of Accelerators and Nuclear Analytical Methods) infrastructure LM 2015056. This publication was supported by OP RDE, MEYS, Czech Republic under the project CANAM OP, CZ.02.1.01/0.0/0.0/16_013/0001812 and by the Czech Science Foundation (GACR No. 19-02482S) and by the University of J.E.Purkyne project UJEP-SGS-2021-53-005-2.

References

1. S. Ramakrishna, *An Introduction to Electrospinning and Nanofibers*, Hackensack, Nj: World Scientific, 2005
2. A. L. Yarin, B. Pourdeyhimi, S. Ramakrishna, *Fundamentals and Applications of Micro- and Nanofibers*, 2014
3. L. Persano, A. Camposeo, D. Pisignano, *Active polymer nanofibers for photonics, electronics, energy generation and micromechanics*, Progress in Polymer Science, 2015, 43, 48-95
4. Y. Gao, F. Guo, P. Cao, et al. *Winding-Locked Carbon Nanotubes/Polymer nanofibers Helical Yarn for Ultrastretchable Conductor and Strain Sensor*, ACS Nano, 2020, 14(3), 3442-3450
5. M.C. Wintersgill, *Ion implantation in polymers*, Nuclear Instruments and Methods in Physics Research Section B: Beam Interactions with Materials and Atoms, 1984, 1(2-3), 595-598
6. P. Ryšánek, P. Čapková, J. Štojdl and col., *Stability of antibacterial modification of nanofibrous PA6/DTAB membrane during air filtration*, Material Science and Engineering, 2019, 96, 807-813
7. M. Mayer, SIMNRA User's guide, Report IPP 9/113, Max-Planck-Institut fur Plasmaphysik, Garching, Germany, 1997.
8. J.F. Ziegler, et al., SRIM: The Stopping and Range of Ions in Matter, Version SRIM2013, Available at: http://www.srim.org/.
9. A. E.Al-Rawajfeh, *Polyamide-based composite membranes: Part 1. Preparation and characterization*, Desalination, 2005, 179(1-3), 265-272
10. Q. Li, Z. Xu, L. Yu, *Effects of mixed solvents and PVDF types on performances of PVDF microporous membranes*, Journal of Applied Polymer Science, 2010, 115(4), 2277-2287
11. J. Gao, J. F. Rusling, D. Zhou, *Carbon-Carbon Bond Formation by Electrochemical Catalysis in Conductive Microemulsion*, The Journal of Organic Chemistry, 1996, 61(17), 5972-5977
12. X. Zhang, M. Renbo, L. Jiang, W. Wei, *Morphology and electrical properties of polypropylene/polyamide 6/glass fibers composites with low carbon black loading*, Journal of Polymer Engineering, 2019, 39(9), 813-821

An Optimized D-T neutron Generator Shielding for Prompt Gamma Ray Neutron Activation Analysis of Light Elements

Radim Uhlář[1,*] and *Petr* Alexa[1,**]

[1]Department of Physics, Faculty of Electrical Engineering and Computer Science, VŠB-Technical University of Ostrava, 17. listopadu 2172/15, CZ-708 00 Ostrava, Czech Republic

Abstract. We present an MCNP analysis of optimal gamma-detector shielding materials and their thicknesses for 14 MeV D-T neutron generator for detection of light elements by prompt gamma ray neutron activation analysis. Tungsten seems to be the best choice for the primary neutron-beam shielding of a minimum thickness, but copper and iron represent an economically acceptable alternative. To shield the gamma detector against gamma rays produced from neutron interactions in the laboratory walls another shielding made from lead is necessary.

1 Introduction

Compact neutron generators are nowadays used in a growing number of applications such as non-destructive elemental analysis of various types of materials including CBRNE materials (chemical, biological, radiological, nuclear and explosives) or drugs. Several neutron based methods can detect explosives that contain four basic elements (C, H, N, O), such as Thermal Neutron Analysis, Fast Neutron Analysis, Pulsed Fast Neutron Analysis and Pulsed Fast/Thermal Neutron Analysis [1]. The sensitivity of the method is closely linked to the reduction of detector background caused by primary or secondary neutrons and gamma rays emitted by surrounding materials. We present an MCNP analysis of optimal neutron and gamma-ray shielding materials and their thicknesses for a HPGe gamma spectrometer in the field of a 14 MeV D-T neutron generator.

2 Materials and Methods

In the Laboratory of Neutron Activation Analysis of VŠB-TU Ostrava we operate a D-T neutron generator (MP320, Thermo Scientific) and a HPGe gamma spectrometer (GC-3018, Canberra). The gamma ray background is due to interaction of neutrons with materials of the neutron generator (NG), gamma spectrometer (GS), laboratory walls and shielding. To minimize the signal-to-background ratio, it is necessary to:

*Corresponding author: radim.uhlar@vsb.cz
**Corresponding author: petr.alexa@vsb.cz

1. Test the ability of materials to shield neutron radiation at the location of the gammaspectrometer as a function of the shielding thickness neglecting the environment (e.g. the laboratory walls);

2. Design the final shielding geometry including the sample, the detector and the environment based on the simulation of the detector response.

The general-purpose 3-D Monte Carlo N-Particle Code (MCNP6) [2] was used to simulate radiation transport. For the MCNP6 simulations, continuous-energy neutron data libraries ENDF71x were used [3, 4]. The NG was modelled as a point non-isotropic D-T neutron source housed in a sulfur hexafluoride-filled aluminium housing (Fig. 1, for more details see [5]). Materials not containing the basic elements of explosives were first selected to be tested as candidates for the final shielding composition [6] (Al, AlF_3, Cu, Fe, Fluental, MgF_2, Mo, Pb, PbF_2, PbF_4, TiF_3, V, W, Zr_3Al).

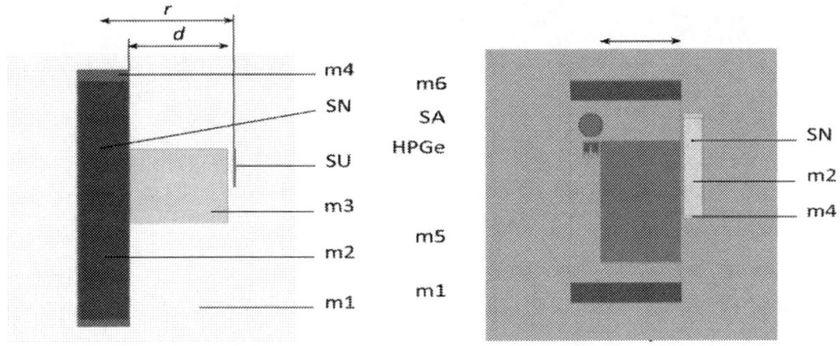

Figure 1. Geometry of the shielding used in the test of different materials neglecting the environment (left) and for the design of the final shielding geometry including the environment (right). Notation: m1 – air, m2 – sulfur hexafluoride, m3 – shielding, SU – testing surface, SN – point source of neutrons, m4 – aluminium, m5 – iron, m6 – lead (it surrounds the neutron shielding, detector and sample at both front and back sides as it is indicated by the thin lines), HPGe – High Purity Germanium detector GC-3018, SA – sample; r is the distance between the center of the testing surface and the neutron source and d and d_1 are the shielding thicknesses.

For thicknesses 10-40 cm with a step of 10 cm, neutron and photon fluxes (tally F2, 9×10^8 histories) in the testing surface (SU) behind the shielding (Fig. 1, left) were calculated. For the five most effective neutron shielding materials, the neutron and photon fluxes were calculated also for thicknesses 5-50 cm (step 5 cm) and 60-90 cm (step 10 cm). In this case a sandwiched structure of the shielding made of different combinations of the five materials was investigated. The results were also compared with neutron shielding efficiency of polyethylene (PE), borated polyethylene (BPE, mass fraction of B: 5%) and lithium polyethylene (LPE, mass fraction of Li: 12%).

To apply these results to our D-T neutron generator placed in the center of a rather narrow laboratory hall (width 2.4 m, length 3.8 m, height 2.7 m), we have to take into account neutron reflection and gamma rays emitted by surrounding materials (laboratory walls). In the MCNP6 simulations as a tested sample we chose 1 kg urea containing all light elements (C, H, N, O) we would like to detect. The sample was positioned close to the detector endcap. Both sample and the HPGe detector were placed in the lead shielding of different thicknesses from 0 cm to 40 cm (step 10 cm) with openings left for the primary neutron beam irradiating the sample (Fig. 1, right). The geometry of the laboratory hall including its walls was taken into account in the MCNP6 simulations [7].

3 Results

The selected criterion K of the neutron shielding efficiency to be minimized, $K = r^2\phi$, accounts for the spherical attenuation of the neutron flux (ϕ is the overall neutron flux per source neutron evaluated by the tally F2 and r represents the distance between the center of the testing surface and the source of the neutrons).

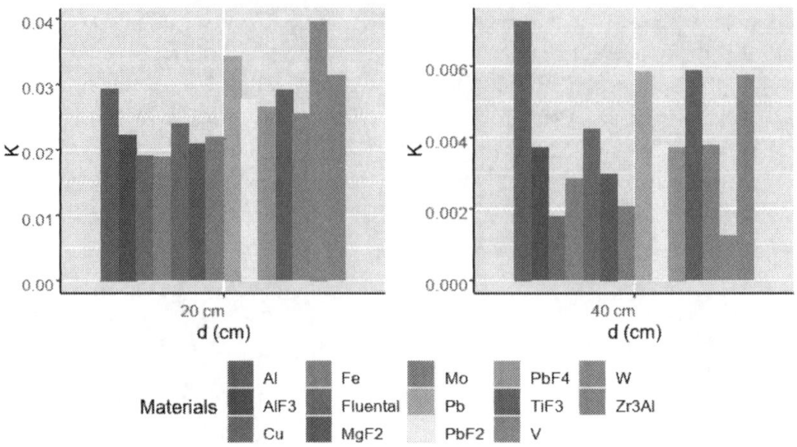

Figure 2. The criterion K (per source neutron) calculated for two thicknesses of different shielding materials.

Fig. 2 shows that the shielding thickness of 20 cm is not sufficient, the thickness of 40 cm gives better results, especially for the following materials that were selected for the more detailed shielding testing: Cu, Fe, MgF_2, Mo, and W. Table 1 shows best sandwiched shieldings in the range of thicknesses from 5 cm to 90 cm.

Table 1. Material composition and thicknesses of the shielding layers minimizing the criterion K.

d (cm)	Material Composition	K	d (cm)	Material Composition	K
5	MgF_2	0.059	40	W (35 cm) Mo (5 cm)	0.0013
10	Fe	0.045	45	W	0.0008
15	Fe (10 cm) MgF_2 (5 cm)	0.028	50	W	0.0006
20	Fe (10 cm) MgF_2 (10 cm)	0.018	60	W	0.0005
25	W	0.009	70	Mo	0.0005
30	W (25 cm) MgF_2 (5 cm)	0.004	80	Mo (70 cm) W (10 cm)	0.0005
35	W	0.002	90	MgF_2	0.0005

The criterion K for the best heavy (W) and light materials (BPE) minimizing the criterion K, and Fe and Cu is presented in Fig. 3. Benefit from a further increase of the shielding thickness beyond 40 cm for heavy materials and beyond 60 cm for light materials is very small due to the criterion K saturation. If we change, e.g., the tungsten shielding thickness from 35 cm to 40 cm, the criterion K decreases by a factor of 1.8, while for the change from 40 cm to 45 cm it decreases 1.5 times only. Similarly, for BPE a thickness increase from 50 cm to 60 cm represents a decrease of K by a factor of 2.0, while for the change from 60 cm to 70 cm K decreases 1.6 times.

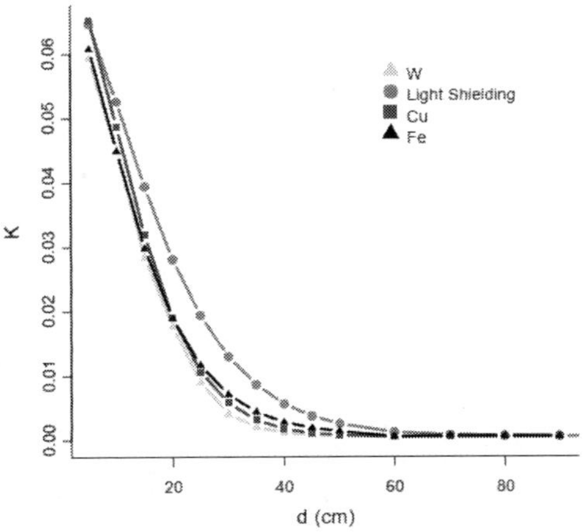

Figure 3. The criterion K as a function of the thickness of the best heavy (W), the best light (BPE), and Cu and Fe shieldings.

The shielding itself is activated by neutrons and is, therefore, a source of gamma radiation which mainly concerns the light shielding based on polyethylene (Fig. 4). The large gamma peak at 4.44 MeV from the inelastic neutron scattering on ^{12}C from PE, BPE or LPE practically rules out carbon detection.

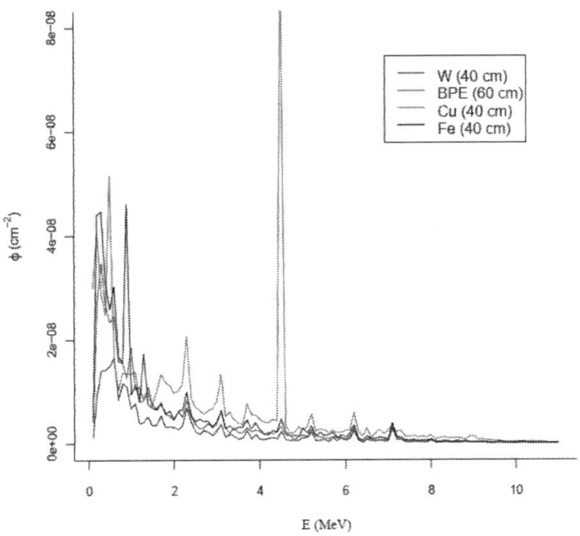

Figure 4. Spectrum of the photon flux per source neutron (tally F2) computed for the surface SU for the best heavy (W), the best light (BPE), and Cu and Fe shieldings.

For the MCNP6 simulations of the effect of surrounding materials at our laboratory on the gamma *spectrum* of the urea measured by the HPGe spectrometer, the 40 cm Fe shielding of the primary neutron beam irradiating the gamma spectrometer as an economically affordable option was chosen. Using tally F8, we found that at least a 10 cm lead shielding of the HPGe gamma spectrometer is than necessary to detect gamma peaks at 2.22 MeV, 4.44 MeV, and 6.13 MeV as prompt gamma rays identifying H, C, and O (see Table 2).

Table 2. Urea + background signal $\phi(E)_{tot}$ relatively to the background signal $\phi(E)_{bg}$ for H, C, and O peaks for two different thicknesses of the lead shielding.

Element	Energy (MeV)	$\phi(E)_{tot}/\phi(E)_{bg}$ Pb 0 cm	$\phi(E)_{tot}/\phi(E)_{bg}$ Pb 10 cm
H	2.22	0.99	1.21
C	4.44	4.1	6.8
O	6.13	1.06	3.7

In the MCNP6 simulations, the fractional standard deviation (relative error) for total flux (tally F2) is less than 0.2%. For the simulated gamma spectra using the pulse height tally F8 at individual energy intervals of the length of approximately 0.7 keV and the thickness of the Fe shielding of 40 cm, the relative error reaches tens of percent for energies of the photons higher than 1 MeV and significant peaks values.

4 Conclusions

We found that heavy shielding materials are more efficient than the light ones because one gets the optimal heavy shielding thickness around 40 cm while for the light shielding it lies around 60 cm. Moreover, the light shielding produces gamma peaks of C and H that we want to detect in investigated samples. The best shielding material (W) is too expensive, Fe or Cu represent economically more acceptable alternatives. Therefore, for the final shielding construction in the laboratory at VŠB-TU Ostrava 40 cm Fe shielding of the primary neutron beam and 10 cm lead shielding of the HPGe spectrometer have been chosen.

This work was supported by the Ministry of Education, Youth and Sports of the Czech Republic through the eINFRA CZ (ID:90140) and by the Ministry of the Interior of the Czech Republic under the project Application of radiation methods for detection and identification of CBRNE materials (VI20192022171).

References

1. IAEA, *Neutron generators for analytical purposes. IAEA Radiation Technology Reports No.1* (IAEA, Vienna, Austria, 2012)
2. *MCNP Users Manual - Code Version 6.2. report LA-UR-17-29981* (Los Alamos National Laboratory, 2017)
3. J. L. Conlin, S. J. Gardiner, D. K. Parsons, A.C. Kahler, B. Lee, and M.C. White, *Release of ENDF/B-VII.1-based Continuous Energy Neutron Cross Section Data Tables for MCNP. Report LA-UR-20240* (Los Alamos National Laboratory, 2013)
4. J. L. Conlin, D. K. Parsons, S. J. Gardiner, M. G. Gray, B. Lee, A.C. Kahler, and M.C. White, *Continuous Energy Neutron Cross Section Data Tables Based upon ENDF/B-VII.1. Report LA-UR-13-20137* (Los Alamos National Laboratory, 2013)
5. R. Uhlář, P. Alexa, O. Harkut, and P. Haroková, IEEE Trans. Nucl. Sci. 67, 382–388 (2020)
6. R. Uhlář, M. Kadulová, P. Alexa, and J. Pištora, J. Radioanal. Nucl. Chem. 300, 809–818 (2014)
7. R. Uhlář and P. Alexa, Radiat. Prot. Dosim. 185, 116-123 (2019)

Effect of Using Different Data Libraries and Simulation Codes on the Calculation of Spectra and Operational Quantities for the D_2O-^{252}Cf Source at PTB

Amer A. Al Qaaod[1*], *Marcel* Reginatto[1], *Andreas* Zimbal[1], and *Miroslav* Zbořil[1]

[1]Physikalisch-Technische Bundesanstalt (PTB), Bundesallee 100, 38116 Braunschweig, Germany

Abstract. The neutron reference field produced by a heavy-water moderated ^{252}Cf source is used at PTB for calibrating neutron-measuring devices. Knowledge of the precise neutron spectrum is very important for the investigation of operational radiation protection quantities. Recently, new fission spectrum data of ^{252}Cf has been proposed based on the latest nuclear data library version. At PTB, earlier calculations for the D_2O moderated ^{252}Cf neutron source were carried out more than 20 years ago, thus updated and more detailed calculations are required. In this paper, a detailed simulation model of the PTB moderated ^{252}Cf source assembly has been prepared and investigated using new spectral data and two different Monte Carlo transport codes MCNP6.1 and PHITS3.22, with ENDF/BVIII.0, ENDF/B-VII.1, ENDF/B-V, and ENDL85 evaluated nuclear data libraries. The results show that the evaluated nuclear data libraries influence the calculated operational quantities by (3-5) %. The dosimetric quantities calculated with the PHITS code and the ENDF/B-VII.1 data library agree well with the MCNP6 results.

1 Introduction

The increasing use of radiation and neutron sources for research and industry applications makes it necessary to calibrate neutron measuring devices in well-characterized calibration fields [1], [2]. The International Organization for Standardization (ISO) recommends neutron reference radiations for calibrating neutron measuring devices [3], [4]. For routine calibrations, different types of radioactive neutron sources with wide-ranging neutron energy spectra should be used. An irradiation facility with ISO recommended radioactive neutron sources such as ^{252}Cf, unmoderated and moderated by heavy water (D_2O), and ^{241}Am-Be, is available at the Physikalisch-Technische Bundesanstalt (PTB), the National Metrology Institute of Germany, for calibrations of area monitors and personal neutron dosimeters [5]– [7].

* Corresponding author: amer.al-qaaod.ext@ptb.de

Earlier Monte Carlo simulations were performed at PTB for the irradiation facility some 20 years ago [5], [6]. In that study, a simple model for the facility with radionuclide neutron sources, D_2O moderated ^{252}Cf, ^{252}Cf bare and ^{241}Am-Be sources, was constructed to investigate the influence of the wall, the air, and the shadow object on the neutron fluence rate at the measurement position. Also, the different spectra were calculated and measured using a Bonner Sphere Spectrometer (BSS) [8]. The authors used a Maxwellian distribution with spectrum parameter T = 1.42 MeV to describe the probability density for a single neutron emission from the ^{252}Cf source. The Monte Carlo N-particle code (MCNP) version 4A with the ENDL85 cross section data library was used to calculate and evaluate the operational radiation protection quantities.

In this paper, a recent Monte Carlo calculation of the D_2O moderated ^{252}Cf neutron reference assembly at PTB is presented and discussed, taking into account the updated fission spectrum data [9], cross section library, more details of the source configuration, and using two different transport codes, MCNP6.1 [10] and the Particle and Heavy Ions Transport code System (PHITS) version 3.22 [11].

2 The D_2O- ^{252}Cf assembly description and simulation model

At PTB, the D_2O-^{252}Cf assembly consists of a ^{252}Cf source placed in a cylindrical capsule of height 11.3 mm and outer surface diameter 10.8 mm made of stainless steel. The source capsule is guided to the centre of the assembly with a stainless-steel tube of wall thickness 1 mm. The heavy water is enclosed in a spherical shell of stainless-steel thickness 1 mm and surrounded by a Cadmium (Cd) half-shell. The accurate diameter of the D_2O container is 299.2 mm, this value was used in the present calculations. The geometrical model is shown in Fig. 1.

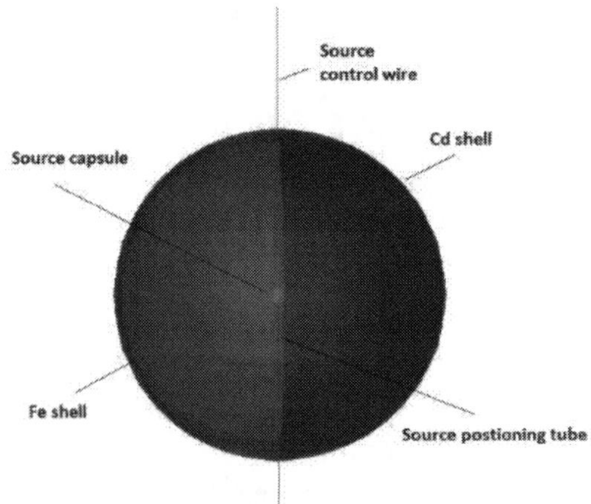

Fig. 1. 3D MCNP6 geometrical model of the D_2O-^{252}Cf assembly at PTB.

The codes MCNP version 6.1 and PHITS version 3.22 were used with the cross-section data libraries ENDF/B-VIII.0, ENDF/B-VII.1, ENDF/B-V, and ENDL85. The codes were run using a High-Performance Computing (HPC) Cluster in Berlin, which is a key facility for centralized scientific computing at PTB. Use of the cluster reduced the calculation time and allowed for a reduction in the uncertainty. The number of source-particle histories run in the problem was 10^9 so the relative error was kept under 0.07 %.

The materials composition of the assembly was described in detail in the model using values obtained from the "Compendium of Material Composition" [12]. Thermal neutron scattering treatment for the heavy water and iron at 293.6 K were considered in the material card in both codes. The D_2O-^{252}Cf assembly model assumes a free-space field, meaning that the source and moderator configuration are surrounded by vacuum (no air, walls, and equipment), and the spectral fluence for the direct neutron component was scored at 100 cm from the geometrical centre of the source at the Cd side using a point detector tally F5.

3 Results and discussion

The ENDF/B-VIII.0 neutron data which is used in the present investigation was released in 2018 [13]. It provides cross-section data for 557 nuclides and thermal scattering libraries for 33 different compounds in ENDF-6 files, whereas the earlier ENDF/B-VII.1 provides cross-section data for 423 nuclides. Also, the ENDF/B-VIII.0 library, in contrast to ENDF/BVII.1, has major changes for neutron reactions for the major actinides and other nuclides that impact simulations. The important isotopes ^{1}H, ^{16}O, ^{56}Fe, ^{235}U, ^{238}U, and ^{239}Pu have been the focus of the international CIELO collaboration and the resulting advances have been incorporated into ENDF/B-VIII.0 [14]. As an example of changes, we point out a previous renormalization of the (n,α) cross section downwards by 32 % for ENDF/B-VII which has now been removed in the new data library [13].

The aforementioned isotopes, specifically ^{16}O and ^{56}Fe, are major components of the D_2O^{252}-Cf assembly (see section 2) and the update of their cross-section data would be expected to influence the dosimetric parameters calculation. This is leading us to re-evaluate those parameters using different release cross section data libraries and discuss their impact. Another point to be taken into account is that the use of different codes in the analysis of a calibration problem is both an essential feature and a source of strong feedback for further improvements in data through mutual comparison of results.

For this purpose, MCNP6.1 and PHITS3.22 were used with four evaluated nuclear data libraries, ENDF/B-VIII.0, VII.1, V, and ENDL-85, to calculate the quantities of interest for the realistic model at PTB. The PHITS code used to simulate the D_2O-^{252}Cf assembly model incorporated the same dimensions, material densities, fission spectrum data and fluence-todose conversion coefficient used for the MCNP6 code. Tally [t-point] was chosen to score the fluence rate at 100 cm from the geometrical centre on the Cd side, which corresponds to the F5 tally in MCNP. For the neutron energy below 20 MeV, the PHITS code makes use of the cross-section data library when the option of using data library is turned on.

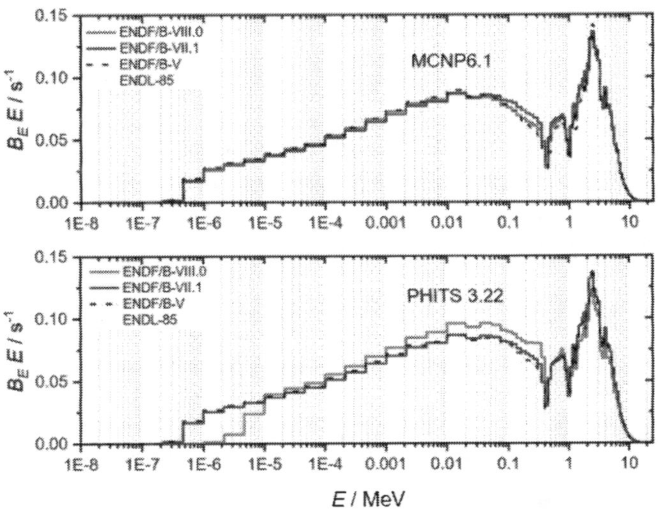

Fig. 2. A comparison of neutron spectra of the D_2O-^{252}Cf source assembly at 100 cm distance in vacuum for different released cross section data libraries, using the MCNP6.1 and PHITS3.22 codes.

The neutron spectrum for the assembly at 100 cm is shown in Fig. 2 for both codes and different cross section data. The MCNP6.1 calculation result shows a slight difference between the evaluated nuclear data libraries used in this comparison, especially in the case of ENDF/B-VII.1 in the energy range from 70 keV to 2 MeV. Also, one can observe a discrepancy in the dosimetric parameter values in Table 1, the ambient $h^{*}(10)$ and personal h_p (10, 0) conversion coefficient in the case of ENDL85 and ENDF/B-V are 108.34, 113.14 and 110.30 and 115.17 pSv cm^2 respectively, lower than the ENDF/B-VIII.0 values which are 113.78 and 118.81 pSv cm^2 respectively. Moreover, the discrepancy between ENDF/BVIII.0 and ENDF/B-VII.1 for calculated ambient and personal conversion coefficient was about 3 %, this is due to the higher values of the spectrum in the energy range between 0.07 and 2 MeV when using ENDF/B-VII.1 with MCNP6, as shown in Fig. 2 (top).

The ENDF/B-VIII.0 showed large deviation compared to the other evaluated nuclear data libraries when used with the PHITS code, the differences between the two codes MCNP6.1 and PHITS3.22 with same data library ENDF/B-VIII.0, ENDF/B-VII.1, ENDF/B-V, and ENDL85 are presented in Fig. 3a, b, c, and d respectively. The neutron spectrum calculated by PHITS code matches the MCNP6 spectrum when the ENDF/B-VII.1 and ENDL85 nuclear data library are used as shown in Figs. 3b and 3d, also the dosimetric parameters are in good agreement as can be seen in Table 1. For the ENDF/B-VIII.0 library the dose quantities are very similar in both codes although the fluence values and the spectral shapes are different. This is due to the dependence of the ambient and personal dose conversion coefficient on the neutron energy. As we can see from Figure 3a, the fluence for energies lower than 0.4 MeV is much higher when using PHITS than MCNP6.1, but these neutrons do not make a significant contribution to the dose quantities as do neutrons with energies greater than 1 MeV.

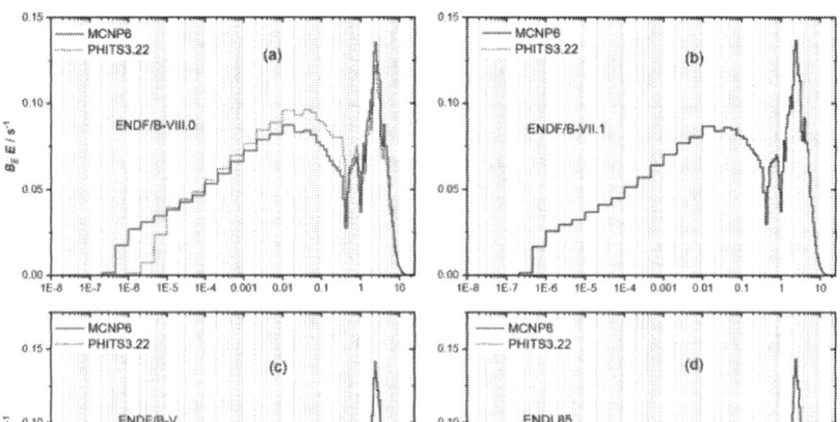

Fig. 3. Neutron spectra of the D_2O-^{252}Cf assembly at 100 cm distance calculated with MCNP6.1 and PHITS3.22 using four different data libraries (a) ENDF/B-VIII.0, (b) ENDF/B-VII.1, (c) ENDF/B-V and (d) ENDL-85.

Table 1. The fluence rate (Φ), fluence-average energy ($\bar{E}_{\Phi}$), ambient dose-average ($\bar{E}_H$), ambient $h^*(10)$ and personal h_p (10, 0) conversion coefficient at 100 cm from the source, variation with cross section data libraries for the D_2O-^{252}Cf assembly of PTB using the codes MCNP6 and PHITS. The relative uncertainty in the fluence rate and dose quantities was below 0.07 %.

Data library	Code	Φ /cm^{-2}	$\bar{E}_{\Phi}$ / MeV	$\bar{E}_H$ / MeV	$h^*(10)$ / pSv cm^2	h_p (10, 0) / pSv cm^2
ENDF/B-VIII.0	*MCNP*	6.99×10^{-6}	0.57	1.99	113.78	118.81
	PHITS	7.76×10^{-6}	0.58	1.98	113.71	118.66
ENDF/B-VII.1	*MCNP*	7.02×10^{-6}	0.58	1.98	117.20	122.44
	PHITS	7.00×10^{-6}	0.58	1.98	117.86	123.01
ENDF/B-V	*MCNP*	6.99×10^{-6}	0.56	2.04	110.30	115.17
	PHITS	7.00×10^{-6}	0.58	2.00	115.28	120.33
ENDL-85	*MCNP*	6.97×10^{-6}	0.55	2.04	108.34	113.14
	PHITS	6.96×10^{-6}	0.55	2.04	108.52	113.33

4 Conclusions

In this paper, a numerical investigation of the heavy water moderated ^{252}Cf source at PTB was carried out taking into consideration the latest fission spectrum data of ^{252}Cf, different evaluated nuclear data libraries and two recent versions of simulation codes, MCNP6.1 and PHITS3.22. The results of the calculation showed that the evaluated nuclear data libraries influence the calculated ambient and personal dose conversion coefficient for the D_2O-^{252}Cf assembly, and a significant difference in the calculated parameters between the latest released ENDF/B-VIII.0 and ENDF/B-VII.1, ENDF/B-V and ENDL85 has been found when the calculations are done with the same code MCNP6.1. ENDF/B-VIII.0 does not work properly with the PHITS code, a large deviation in the neutron spectrum and fluence rate with MCNP6.1 has been reported, and further validation should be performed for ENDF/B-VIII.0 considering single isotopes to figure out the source of the discrepancy. In general, the PHITS code proved its reliability and accuracy to simulate a heavy water moderated ^{252}Cf neutron source using ENDF/B-VII.1, where the results agree well with MCNP6.1 which contains the same cross section data library.

In future work, a comparison between the dosimetric quantities of the ISO and PTB moderator model using the previous (ISO 8529-1 2001) and a new proposed ^{252}Cf fission spectrum data (ISO 8529-1 2021) will be discussed. Furthermore, a realistic calculation including details of the calibration facility at PTB will be performed and evaluated taking into consideration the room return neutron components.

This work was supported by the Philipp Schwartz Initiative of the Alexander von Humboldt Foundation and the Federal Foreign Office of Germany which made available a scholarship for Amer Al-Qaaod. Special thanks to Désirée Radeck for formulating and guiding the work of this project and for her crucial contributions to the results. We thank Sebastian Fässer, Sven Koch, and Sebastian Reinert for assistance with the calibration facility. Our thanks to all staff in the HPC in PTB Berlin for their support regarding the use of the computer cluster.

References

1. V. Gressier, V. Lacoste, A. Martin, and M. Pepino, "Characterization of a measurement reference standard and neutron fluence determination method in IRSN monoenergetic neutron fields", Metrologia, 51, no. 5, p.431, 2014.
2. M. Ginjaume, "Active methods & instruments for personal dosimetry of external radiation : present situation in Europe and future needs," *Current trends in radiation protection,* EDP Sciences, 65–82, 2021.
3. J. Böhm *et al.*, "ISO recommended reference radiations for the calibration and proficiency testing of dosemeters and dose rate meters used in radiation protection," *Radiat. Prot. Dosimetry*, vol. 86, no. 2, pp. 87–105, 1999.
4. D. Thomas, R. Bedogni, R. Méndez, A. Thompson, and A. Zimbal, "Revision of ISO 8529-Reference neutron radiations," *Radiat. Prot. Dosimetry*, vol. 180, no. 1–4, pp. 21–24, 2018.
5. H. Kluge, "Irradiation facility with radioactive refernce neutron sources: Basic principles," *PTB-Report*, PTB-N-34, 1998.
6. S. Jetzke and H. Kluge, "Characteristics of the ^{252}Cf neutron fields in the irradiation facility of the PTB," *Radiat. Prot. Dosimetry*, vol. 69, no. 4, pp. 247–256, 1997.
7. H. Kluge, W. Rasp, J.B. Hunt, "An intercomparison of the D_2O-moderated ^{252}Cf reference neutron sources at PTB and NPL," *Radiat. Prot. Dosimetry*, vol. 11, no. 1, p. pp.61–63, 1985, [Online]. Available: https://doi.org/10.1093/oxfordjournals.rpd.a079446.
8. H. Kluge, A.V. Alevra, S. Jetzke, K. Knauf, M. Matzke, K. Weise and J. Wittstock, "Scattered neutron reference fields produced by radionuclide sources," *Radiat. Prot. Dosimetry*, vol. 70, no. 1–4, pp. 327–330, 1997.
9. R. Méndez, J. M. Gómez-Ros, D. J. Thomas, A. K. Thompson, and R. Bedogni, "Revision of the ^{252}Cf and D_2O-moderated ^{252}Cf reference neutron fields for use in radiation protection dosimetry," *Radiat. Phys. Chem.*, vol. 184, p. 109433, Jul. 2021, doi: 10.1016/j.radphyschem.2021.109433.
10. T. Goorley *et al.*, "Features of MCNP6," *Ann. Nucl. Energy*, vol. 87, pp. 772–783, 2016, doi: 10.1016/j.anucene.2015.02.020.
11. T. Sato *et al.*, "Features of Particle and Heavy Ion Transport code System (PHITS) version 3.02," *J. Nucl. Sci. Technol.*, 2018, doi: 10.1080/00223131.2017.1419890.
12. R.J. McConn Jr, C.J. Gesh, R.T. Pagh, R.A. Rucker, R.G. Williams III , "Compendium of Material Composition Data for Radiation Transport Modeling," 2011, [Online]. Available: www.pnnl.gov/main/publications/external/technical_reports/pnnl-15870rev1.pdf.
13. D. A. Brown *et al.*, "ENDF/B-VIII.0: The 8th Major Release of the Nuclear Reaction Data Library with CIELO-project Cross Sections, New Standards and Thermal Scattering Data," *Nucl. Data Sheets*, vol. 148, pp. 1–142, 2018, doi: 10.1016/j.nds.2018.02.001.
14. M. B. Chadwick *et al.*, "The CIELO collaboration: neutron reactions on ^{1}H, ^{16}O, ^{56}Fe, ^{235,238}U, and ^{239}Pu," *Nucl. Data Sheets*, vol. 118, p. pp.1–25, 2014, [Online]. Available: https://doi.org/10.1016/j.nds.2014.04.002.

Nanoparticles embedded in a sponge of polydimethylsiloxane by laser ablation in liquid

Mariapompea Cutroneo[1*], *Vladimir* Havranek[1], *Lorenzo* Torrisi[2], *Letteria* Silipigni[2], *Lubomir* Kovacik[3], *Petr* Malinsky[1,4], *Josef* Flaks[1], *Petr* Slepicka[5], *Dominik* Fajstavr[5], *Olga* Janoušková[6], *Daniela* Zbořilová[6] and *Anna* Mackova[1,4]

[1]Nuclear Physics Institute, AS CR, 25068 Rez, Czech Republic
[2]Department of Physics (MIFT), Messina University, V. le F.S. d'Alcontres 31, 98166 S. Agata, Messina, Italy
[3]Center for Cellular Imaging and NanoAnalytics (C-CINA) Biozentrum, University of Basel, CH-4058 Basel, Switzerland
[4]Department of Physics, Faculty of Science, University of J. E. Purkyně, Pasteurova 3544/1, 400 96 Ústí nad Labem, Czech Republic
[5]Department of Solid State Engineering, Institute of Chemical Technology, 166 28 Prague, Czech Republic
[6] Department of Biology, Faculty of Science, University of J. E. Purkyně, Pasteurova 15, 400 96 Ústí nad Labem, Czech Republic

Abstract. This work describes the preparation of polydimethylsiloxane (PDMS) sponge with pore sizes of about 50 and 900 μm. The sponges synthetized by the sugar template process were embedded with graphene oxide (GO) and gold nanoparticles (AuNPs) previously produced by laser ablation in liquid. The suspension containing graphene oxide and gold nanoparticles were optically characterized by UV-ViS spectroscopy. The dispersion of the nanoparticles in the PDMS sponges was observed by the Scanning Electron Microscopy (SEM). The biocompatibility of virgin PDMS, PDMS filled with graphene oxide, and with graphene oxide and gold nanoparticles was studied for different types of cell cultures. This study has allowed us to confirm that the PDMS sponge is a good matrix for embedding AuNPs and has highlighted as the presence of GO hinders the aggregation of AuNPs avoiding the use of surfactant and allowing their use in biological applications.

1 Introduction

Polymer composites have recently attracted intensive research interest in nanoelectronics, polymeric bionanomaterials, nanocomposite-based drug delivery systems, etc. The loading of fillers in polymers can produce a significant improvement in mechanical, thermal, electrical, optical, properties. The functionalization of the polymer obtained by incorporating nanoparticles (NPs), is correlated to the NPs homogeneous distribution and

* Corresponding author: cutroneo@ujf.cas.cz

dispersion. In this frame, the use of porous 3D material appears promising for larger surface area-to-volume ratio, easier mixing of filler whatever the wettability of the native polymer is, better anchoring of filler and synthesis of the final material. In our previous papers **[1]** polydimethylsiloxane (PDMS) sponges with porosity of 50-900 μm were produced for applications in biomedicine **[2]**. PDMS is an elastomer containing the Si-O-Si backbone which is responsible for its flexibility, handling, electrical resistance **[3]**. Further characteristics making PDMS promising for biomedical applications such as cell-based implants are represented by: chemical inertness, oxygen permeability, thermal stability, mechanical resistance. The desired structural, surface and functional modifications of the PDMS are explicitly related to the type, amount and choice of fillers. Metals **[4]**, nanoclays, carbon nanotubes and graphene are mostly applied for improving strength, antimicrobial resistance, adhesion, electrical and thermal conductivity, multi-cyclic deformations, stress endurance for advances in polymer composites. The presence of gold nanoparticles (Au NPs) is evinced when an incident light beam is strongly absorbed by the nanostructures at characteristic wavelength bands. This is a consequence of the coupling of the electric field of the incident light to the electric-field induced dipolar oscillations of the valence electrons on the surface of metallic NPs **[5]**.

Another promising nanosized material used as filler, is graphene oxide (GO). It is a single layer of graphite with the epoxide and hydroxyl groups at the basal plane and the carbonyl and carboxyl groups at the edges. The physical properties of GO depend on the sp^2 to sp^3 hybridized carbon atoms ratio and on the different functional groups. The native insulating nature of GO can be turned in more conductive reduced graphene oxide (rGO) **[6]** by thermal treatment, laser irradiation, electron and ion irradiation **[7]**. Graphene oxide has been used as antimicrobial agent for the water disinfection to remove organic molecules and waterborne pathogens **[8]**, radiation sensitive material **[9]**, dosimeter **[10]** and sensors **[11]**.

In this work, nanoparticles are produced by laser ablation in medium by monitoring the laser fluence, the laser wavelength, the irradiation time and the type of employed liquid **[12]**. The laser beam, focused on the surface of a solid target placed in a vessel containing a liquid, generates a plasma plume at the solid-liquid interface. This last expands due to the reaction occurring inside and a vapor layer is formed. The vapor layer expands into a cavitation bubble while the plasma plume shrinks because of the compression provoked by the vapor layer and then collapses. Nanoparticles are released first in the cavitation bubble and after its collapsing into water. Nanoparticles stop to grow when they reach the same temperature as the water **[13]**.

In particular, in the present study, porous polydimethylsiloxane **[1]**, embedded with gold nanoparticles and gold nanoparticles anchored to graphene oxide, is obtained. Both nanoparticles are produced by laser ablation in liquid and the validity of the method for the preparation of composites for biomedical applications will be reported.

2 Materials and Methods

2.1 Preparation of gold nanoparticles by laser ablation

A nanosecond Nd:YAG Litron laser **[14]** operating at 1064 nm, 4 ns pulse duration, 375 mJ pulse energy, in the 2.5 Hz repetition mode, was employed to irradiate a solid target of gold (99.99 % purity, Goodfellow). The laser beam was directed to a quartz prism and focused by a lens, with a 50 cm focal length, on the Au target placed in a beaker vessel containing 5 ml of liquid. The used setup for the nanoparticle synthesis is shown in **Fig. 1**. Two liquids have been used: 5 ml of distilled water for the production of Au+H_2O and 0.5 ml of GO mixed with 4.5 ml of distilled water for the production of GO+Au+H_2O. The concentration of produced Au NPs was evaluated by weighing the gold target before and after 5 minutes of laser ablation and calculating its weight loss. In accordance with RBS and EDX analyses reported in the following sections, it is reasonable to assume that Au NPs have spherical shape with average diameter of about 10 nm as confirmed by SEM and optical analyses and same density as bulk gold (i.e. 19.3 g/cm^3). The Au NPs concentration was calculated to be 1.0 x 10^{14} NPs/cm^3 in Au +H_2O and 1.9 x 10^{13} NPs/ cm^3 in GO+Au+H_2O. The PDMS (Sylgard 184 Dow Corning) was obtained mixing the prepolymer (monomer) and the curing agent (crosslinker) in a 1:10 weight ratio.

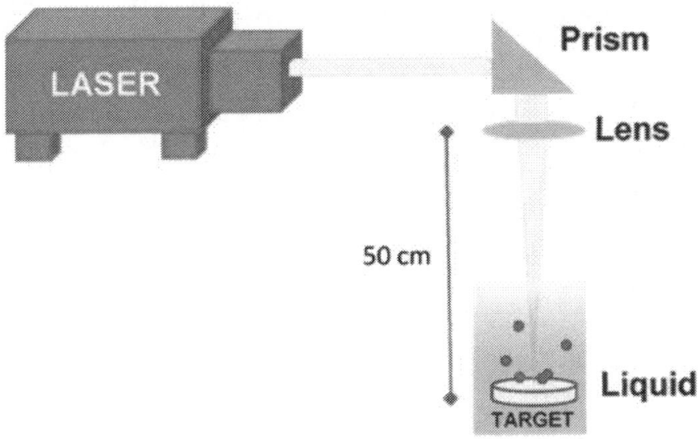

Fig. 1. Sketch of the laser set-up for the synthesis of nanoparticles by laser ablation in medium.

The PDMS solution was placed in a glass beaker containing four sugar bricks (about 1 cm^3 in size) in a vacuum chamber to improve the gassing and the capillarity infiltration of the solution into the hollows of the sugar bricks **[15]**. Successively the sugar bricks filled with the PDMS, were thermally cured in an oven for 45 minutes at 100° C in air. The obtained PDMS sponges were immersed in deionized water at 60° C to completely dissolve the sugar powders and dried in air. Over the four PDMS sponges, three were immersed in glass beakers containing 1.1 g of GO+ H_2O, GO+ Au+H_2O and Au+ H_2O suspensions while the last one was used as virgin for comparison during the following characterization investigations.

2.2 Rutherford backscattering spectrometry (RBS)

The compositional changes of the solutions during the laser irradiation were monitored by Rutherford backscattering spectrometry (RBS). Each one of the three suspensions with a volume of 0.5 ml was poured on silicon substrates and then dried in air. To reduce the background and avoid the generation of resonances in Si during the RBS analysis, a proton beam of 1.3 MeV was chosen. The ion current was 2.2 nA and the beam spot size was (1.8x1.8) mm^2. The backscattered ions were monitored by an Ultra-Ortec PIPS silicon detector located at a 160° scattering angle. RBS energy scale spectra were converted in concentration depth profile using the SIMNRA simulation code **[16]**. The energy loss of protons in the investigated materials has been evaluated using the SRIM code **[17]**.

2.3 UV-Vis spectroscopy

The optical transmittance **[18]** of the suspensions was acquired in the wavelength interval ranging between 300 nm and 700 nm using an AvaSpec-2048 spectrometer with a UB-600 lines/mm grating. The light generated by a halogen-deuterium dual source, illuminated at 0° the front side of the quartz cuvette containing a suspension while the probe connected to the spectrometer was placed behind the cuvette at 180° in the transmission configuration.

2.4 Scanning Electron Microscopy (SEM)

The morphology of the produced nanoparticles in the three suspensions was studied by a TESCAN LYRA3 GMU (Tescan, CZ) system in the secondary-electron mode using an acceleration voltage of 5 kV. The view fields of the probed surfaces were 20 μm, 5 μm and 2 μm for a better identification of the nanoparticles. Being SEM equipped with an Energy Dispersive X-ray spectrocopy (EDX) probe it has been also possible to perform X-ray fluorescence microanalysis induced by (5÷20) keV electron beams on the investigated samples.

2.5 Cytotoxicity analysis

The cytotoxicity analysis was performed using two cancer cell line (human mammary carcinoma cell line-MCF7 and human prostat carcinoma cell line-PC3) and one primary cell line (rat mesenchymal cells – rMSC). The cells were incubated with PDMS composites for 72 hours then the PrestoBlue cell viability assay were used to quantify viability of cells. PrestoBlue is cell permeable resazurin-based solution that functions as a cell viability indicator by using the reducing power of living cells to quantitatively measure the proliferation of cells. Resazurin in living cells is modified by the reducing environment of the viable cell and turns red in colour, becoming highly fluorescent. This color change can be detected using fluorescence or absorbance measurements. Table I lists the types of culture cells as well as the cell media.

Table I - Cell lines, morphology and culture medium of three different culture cells used for cytotoxicity analysis

CELL LINE	PROVENIENCE	MORPHOLOGY	CULTURE MEDIUM
MCF7	human breast/mammary adenocarcinoma	epithelial-like	EMEM
PC3	human prostate adenocarcinoma	epithelial	DMEM/F12
rMSC	rat mesenchymal stem cells	stromal	DMEM

The description of the assay: The cells, $2x10^4$ in 1 ml of cultivation media, were seeded into individual wells of a 24-well plate the day before the addition of prepared samples. The piece of prepared sponge ($2x3$-$4mm^2$) was added to the well with the growing cells and incubated for 72h. As control the cells without sponge and the cells with 5mM/ml H_2O_2 were used. After 72 hours 100 μl of cell viability PrestoBlue reagent was added to the well and incubated at 37°C for 1-3h. The active compound of cell viability assay resorufin is metabolized in viable cells to the resazurin, which has different level of absorbance. Then the medium was used for measurement the absorbance at 560 nm. The experiment was repeated in duplicate three individual times.

3 Results and discussion

Fig. 2 shows the RBS spectra for the GO+ H_2O suspensions poured on Si substrates, dried in air and then analysed using 1.3 MeV H^+ ions at the incidence angle of 60°(a) and at incidence angle 0°(b). RBS has a good depth resolution of the order of several nm, and a very good sensitivity for heavy elements, of the order of parts-per-million (ppm). The analysed depth is typically about tens microns for incident protons. Presently it was not possible to evaluate the thickness of the three solutions after drying because the distributions were not homogeneous and because of the presence of the thick silicon substrate.

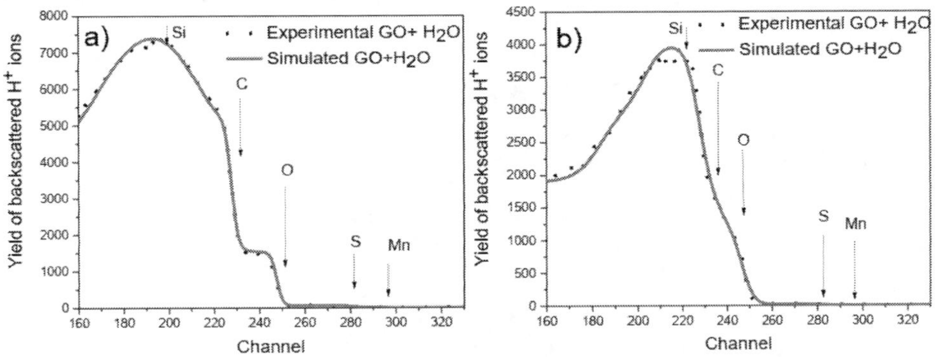

Fig. 2. The RBS spectra for GO+ H_2O suspensions poured on Si substrates, dried in air and then analysed using H^+ ions at the incidence angle of 60°(a) and at the incidence angle of 0°(b).

In Figure 2b the oxygen contribution as well as the carbon one is partially masked by the overlapping of the silicon contribution. At 60° the Si signal is moved to lower energy as expected. It was observed that at the incidence angle of 60° a better discrimination of the oxygen and carbon contributions was observed as reported in Figure 2a. At the 60° incidence angle, a wide fraction of the layer of the suspensions can be investigated.

The detection of the reduction of the oxygen functional groups in GO can be validated by the fragmentation process of the graphene oxide platelets during the laser irradiation of the GO+Au+H_2O suspension which is in agreement with the presented EDX analysis. Table II lists the atomic percentage of the elemental composition of the GO+H_2O, Au+H_2O, GO+Au+H_2O suspensions obtained from the RBS analyses using the 1.3 MeV proton energy.

Table II. Elemental composition of GO+ H_2O, Au+ H_2O, GO+Au+H_2O suspensions poured on silicon substrates evaluated by the RBS analysis.

SAMPLE	Elemental composition of the suspensions (at %)					
	C	O	H	S	Mn	Au
GO+ H_2O	49.0	35.0	15.2	0.3	0.5	-
Au+ H_2O	35.8	28.8	17.2	-	-	18.2
GO+ Au+ H_2O	56.2	25.8	17.0	0.3	0.5	0.2

In Table II the silicon contribution due to the substrate is missing because the evaluation is related the first superficial layers where the silicon contribution was not detected. In GO+ Au+ H_2O is noted a lower Au concentration than in Au+H_2O which is in agreement with the weighing of the gold target before and after the laser irradiation. The ablation rate of gold in GO is lower than in water because of the GO's greatest ability to absorb the laser radiation. The about 26.3 % reduction of the oxygen concentration going from GO+ H_2O to GO+ Au+ H_2O is congruent with the releasing of the oxygen functional group in GO as a consequence of the presence of Au NPs and graphene platelets fragmentation during the laser irradiation **[19]**. In GO+ Au+ H_2O the concentration of gold is a quarter of that in Au+H_2O. The reason is the lower ablation rate in GO than in water as a consequence of the higher attenuation coefficient and absorptivity of GO to the laser than of water. The uncertainty in the evaluation of the elemental concentrations is less than 10 %.

Fig. 3 shows the SEM images with different magnifications of graphene flakes in GO+H_2O a)- b), of the agglomerates of Au nanoparticles in Au +H_2O c)- d) and of the structures in GO+ Au +H_2O e)- f)- g).

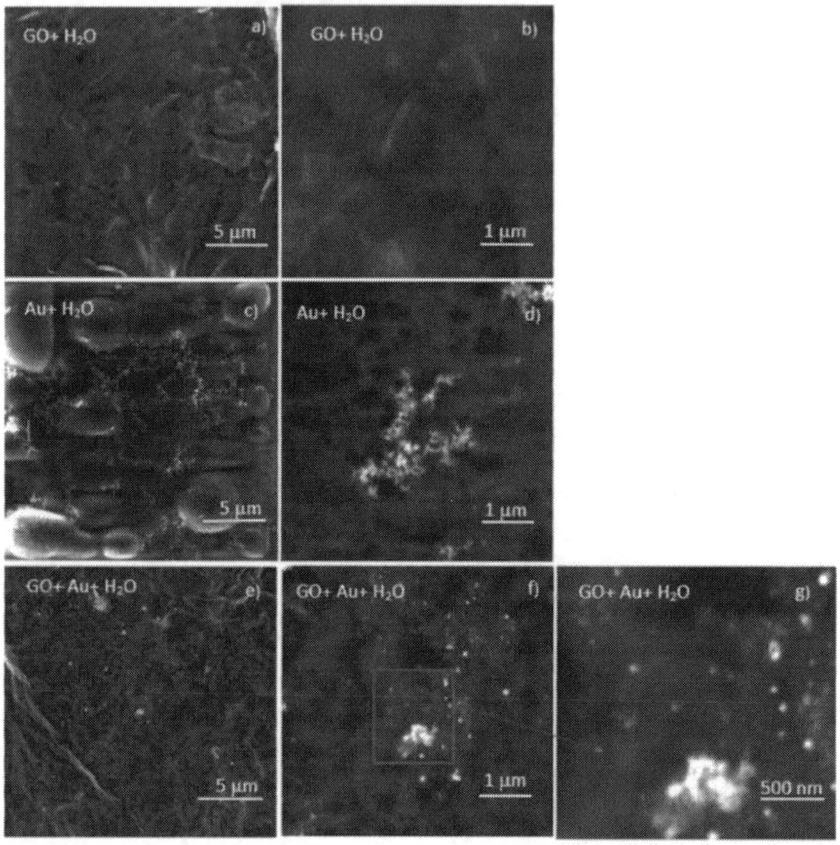

Fig. 3. SEM images of the GO+ Au a) b), Au+ H_2O c), d) and GO+ Au+H_2O e), f), g) deposited on Si substrates at magnifications of 10.4 kx, 41.5 kx and 104 kx.

The SEM view fields were 5 µm, 1 µm and 500 nm while the magnifications were 10.4 kx, 41.5 kx and 104 kx respectively. The concentration of nanoparticles, generated in Au+ H_2O, seems to be higher than in GO+ Au+H_2O in agreement with the RBS and EDX analyses. The size of Au NPs is about 10 nm in Au+H_2O while in GO+ Au+H_2O the bigger size of the structures is ascribable to Au structures anchored to the GO platelets in agreement with the AFM and SEM analyses reported in our previous work **[2]**.

Fig. 4 displays the absorbance as a function of the wavelength, evaluated for the suspensions of GO+ H_2O, GO+ Au+ H_2O and Au+ H_2O immediately after the laser irradiation and after four hours.

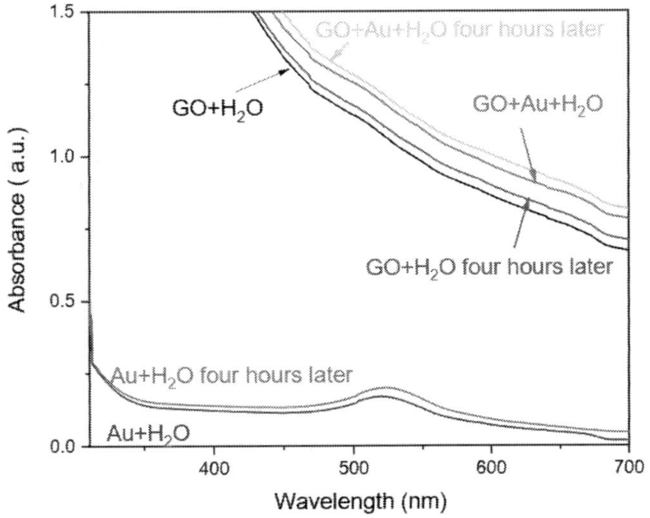

Fig. 4. Absorbance spectra of GO+ H_2O, GO+ Au+H_2O and Au+ H_2O.

In Au+H_2O after four hours of laser ablation production, the SPR (Surface Plasmon Resonance) absorption peak **[20]** around 520 nm is wider due to the Au NPs aggregation. It is in accordance with SEM analysis and assignable to the Au NPs 10 nm average size. The intensity in GO+H_2O increases after four hours from the production likely because of the evaporation of the water content in the suspension. In GO+Au+H_2O the SPR peak is barely visible because of the low attained Au NPs concentration and it is maintained up to four hours after the laser ablation. The intensity of the absorbance in GO+Au+H_2O is higher than in GO+H_2O owing to the presence of reduced GO due to the laser irradiation **[21]** as well as to the presence of Au NPs and to a higher background, due to the light scattering.

4 Biological considerations

The analysis of cytotoxicity of the composites was accomplished monitoring the metabolic cell culture activity as a viability indicator. The active compound of viability assay – Presto blue is resazurin which is reduced to the resorcin in metabolically active cells. The resorcin has different absorbance and fluorescence than resazurin. None of the composites was cytotoxic in any of the used cell cultures. The viability of cells with the composites was not significantly different with viability of nontreated MCF7, PC3 or rMSC cell cultures. This result is interesting for future possible applications because the materials are not cytotoxic to the cancer cell line similarly to the non-cancerous, primary mesenchymal cell line. The virgin porous PDMS was hydrophobic and this explains why it was not wettable in comparison with its two composites - GO+H_2O and GO+Au+H_2O, in which it was modified making its composites hydrophilic. This last aspect is important in biological applications such as the peptide or protein coatings.

5 Conclusion

In the present paper, nanoparticles Au were synthetized by pulsed laser ablation in two different liquids. A nanosecond Nd: YAG laser was focused on the surface of a solid gold target placed either in distilled water or in distilled water containing graphene oxide. In this way two different suspensions were produced: one of Au NPs and the other of Au NPs anchored to GO nanometric platelets. Each one of the suspensions was embedded in porous PDMS sponges.

The RBS analysis showed: 1) reduction of GO during the laser ablation as a consequence of detachment of oxygen functional groups; 2) lower concentration of Au NPs in the suspension of GO+Au+H_2O due to the reduction of ablation rate as a consequence of the high GO absorbance in liquid hindering the interaction between the laser and the solid target.

The production of spherical nanoparticles of about 10 nm in average diameter is indicated from the optical analysis exhibiting an absorption band centred at 520 nm as expected. A bigger size of Au NPs anchored to GO was indicated by the SEM analysis in agreement with the AFM analysis reported in our previous work. The optical analysis suggests that in the suspension containing GO the aggregation of Au NPs doesn't occur confirming the non-need for surfactant and making the suspension clean and without residue from precipitate.

The biological analysis was very positive because it revealed for all the used cells no cytotoxicity risk associated both to the use of the realized composites and the followed manufacture process.

The research has been realized at the CANAM (Center of Accelerators and Nuclear Analytical Methods) infrastructure LM 2015056 and has been supported by project GACR 19-02482S. This publication was supported by OP RDE, MEYS, Czech Republic under the project CANAM OP, CZ.02.1.01/0.0/0.0/16_013/0001812. O. J. and D. Z. acknowledge assistance provided by the Research Infrastructure NanoEnviCz (Project No. LM2018124) and the project Pro-NanoEnviCz (Reg. No. CZ.02.1.01/0.0/0.0/16_013/0001821), supported by the Ministry of Education, Youth and Sports of the Czech Republic and the European Union – European Structural and Investments Funds in the frame of the Operational Programme Research Development and Education, the ERDF/ESF project "UniQSurf - Centre of biointerfaces and hybrid functional materials" (No. CZ.02.1.01/0.0/0.0/17 _048/0007411)

References

1. M. Cutroneo, V. Havranek V. Semian, A. Torrisi, A. Mackova, P. Malinsky, L. Silipigni, P. Slepicka, D. Fajstavr, Journal of Porous Materials **28**, 1481-1491, (2021)
2. M. Cutroneo, V. Havranek, L. Torrisi, L. Silipigni, L. Kovacik, P. Malinsky, P. Slepicka, D. Fajstavr, O. Janoušková, D. Zbořilová, A. Mackova, Vacuum **199**, 110951, (2022)
3. X. M. He, X. J. Mu, Q. Wen, Z. Y. Wen, J. Yang, C. G. Hu and H. F. Shi, Nano Res **9**, 3714-3724 (2016)
4. L. Torrisi, L. Silipigni, N. Restuccia, S. Cuzzocrea, M. Cutroneo, F. Barreca, B. Fazio, G. Di Marco, S. Guglielmino, Journal of Physics and Chemistry of Solids **119**, 62-70 (2018)
5. M. Cutroneo, V. Havranek ,A. Mackova, P. Malinsky,A. Torrisi, L. Silipigni, P. Slepicka, D. Fajstavr, L. Torrisi,et al., Surf Interface Anal., 1–9 (2021)

6. M. Cutroneo, V. Havranek, A. Mackova, P. Malinsky, L. Torrisi, L. Silipigni, B. Fazio, A. Torrisi, K. Szokolova, Z. Sofer, J. Stammers, Materials Chemistry and Physics **232**, 272-277 (2019)
7. M. Cutroneo, L. Torrisi, V. Havranek, A. Mackova, P. Malinsky, A. Torrisi, J. Stammers, Z. Sofer, L. Silipigni, B. Fazio, M. Fazio, R. Böttger, NIMB **460**, 169-174 (2019)
8. R. K. Upadhyay, N. Soin, S. S. Roy, RSC Adv. **4**, 3823–3851 (2014)
9. L. Torrisi, L. Silipigni, M. Cutroneo, A. Torrisi, Vacuum **173** 109175 (2020)
10. L. Torrisi, L. Silipigni, D. Manno, A. Serra, V. Nassisi, M. Cutroneo, A. Torrisi, Vacuum **178** 109451 (2020)
11. L. Torrisi, L. Silipigni and G. Salvato, J. of Materials Sci.: Mater. in Electronics, **31**(14) 11001-11009 (2020).
12. R. Lindley, R. Gilgenbach, C. Ching, J. Lash and G. Doll, Journal of Applied Physics **76**(9), 5457–5472 (1994)
13. G. Yang, Progress in Materials Science **52**(4), 648–698 (2007)
14. https://litron.co.uk/
15. L. Silipigni, G. Salvato, B. Fazio, G. Di Marco, E. Proverbio, M. Cutroneo, A. Torrisi, L. Torrisi, Journal of Materials Science: Materials in Electronics **32**, 8703–8715 (2021)
16. M. Mayer, SIMNRA version 6.06, Max-Planck-Institut fur Plasmaphysik Garching, Germany, 2006. http://www.rzg.mpg.de/~mam/
17. J. Ziegler, M. Ziegler and J. Biersack *NIM B* 268(11-12), 2010, 1818-1823
18. M. Cutroneo, A. Torrisi, V. Ryukhtin, M. Dopita, L. Silipigni, A. Mackova, P. Malinsky, P. Slepicka and L. Torrisi, Journal of Instrumentation **15**, 3, C03044 (2020)
19. G. Moon, Y. Park, W. Kim, and W. Choi, Carbon **49,** 11 3454–3462 (2011)
20. M.A. García, J. Phys. D: Appl. Phys. **44** (28), 283001–281/43 (2011)
21. H. Shi, C. Wang, Z. Sun, Y. Zhou, K. Jin, S. A. T. Redfern and G. Yang, Optics Express **22** (16), (2014)

The multi-energetic Au ion implantation of graphene oxide and polymers

Petr Malinsky[1,2,*], *Josef* Novák[1,2], *Eva* Štěpanovská[1,2], *Petr* Slepička[3], *Václav* Švorčík[3], *Kateřina* Szőkölová[4], *Petr* Marvan[4], *Zdeněk* Sofer[4], *Anna* Mackova[1,2]

[1]Nuclear Physics Institute of the Czech Academy of Sciences, v. v. i., 250 68 Rez, Czech Republic
[2]Department of Physics, Faculty of Science, J.E. Purkinje University, Pasteurova 15, 400 96 Usti nad Labem, Czech Republic
[3]Department of Solid State Engineering, University of Chemistry and Technology, 166 28 Prague, Czech Republic
[4]Department of Inorganic Chemistry, University of Chemistry and Technology, 166 28 Prague, Czech Republic

Abstract. The electric properties of polymers are increasingly important in a wide range of applications such as sensors, energy storages, microelectronics, and filtration membranes among others. In this work, the effect of multi-energetic Au ion implantation on the graphene oxide (GO), polyimide (PI), polyethylene terephthalate (PET) and polylactide (PLLA) elemental, chemical, structural end electric properties is presented with potential application in 3D metal-dielectric structure synthetization. The three energies, 3.2, 1.6, 0.8 MeV of Au ions with fluence 3.75×10^{14} cm^{-2} were used in ascending or descending order to create two sample sets, which were subsequently analysed by RBS, ERDA, EDS and AFM. RBS analysis was used for Au-depth profile characterization in the implanted samples, the profiles agree reasonably with those simulated by SRIM code. Electrical properties were investigated by standard two-point technique with respect to the used parameters of the ion irradiation. The sheet resistance decreases after ion irradiation and it is evident that the ascending order of ion implantation energies has more significant effect on the conductivity enhancement compare to the descending one.

1 Introduction

Fast-growing modernization and industrialization has a negative impact on the growth of the pollution in the air. Thus, the gas sensors are indispensable for the environment and human health protection. The electrochemical resistive types of gas sensors are recently the widely used as sensors and the polymers are huge candidates in sensing applications due to its low cost, commercial availability, and easy deposition on different transducers [1]. Moreover, some usable polymers (e.g. PLLA) are biocompatible and biodegradable and can be used for the fabrication of environmentally friendly sensors. And already, the graphene and its derivatives (especially graphene oxide and reduced graphene oxide) have aroused significant interest for ultrasensitive gas detectors due its high carrier mobility, extraordinary surface area, unique sensing performance, room-temperature working conditions and tremendous application prospects [2,3]. Moreover, the sensitivity of

*Corresponding author: malinsky@ujf.cas.cz

graphene or polymer based sensors can be further enhanced via doping by metal or its oxides or functionalization by ion irradiation [4,5]. The ion irradiation is an efficient technique for the chemical and functional property modification that in case of organic compounds provides possibility of resulting properties selection (degree of reduction, sp^2/sp^3 carbon hybridization ratio modification, catalytic and sensory properties, and dielectric properties). The conductive polymers have important repercussions in the field of flexible electronics, and in sensing technology, and plenty studies of ion-implantation into polymers reported an electrical conductivity enhancement due to polymer carbonization [6]. When the ion fluence exceed 1×10^{13} cm^{-2}, the threshold for individual ion tracks overlapping is overcome and π-bonded carbon clusters grow, aggregate and form a network of conjugated carbon bonds; for the ion fluences above 1×10^{17} cm^{-2} the aggregation phenomena induce the formation of amorphous carbon and/or graphite like material from irradiated polymer [6,7]. Thus, in the range between 5×10^{13} cm^{-2} and 1×10^{17} cm^{-2}, the nucleation and growth of nano-sized carbon-enriched clusters is expected to take place until a quasi-continuous carbonaceous buried layer is formed [6]. The electron hopping or tunnelling between the conductive carbon islands is then considered as the origin of the electrical conductivity of irradiated polymers [8]. The decorating and functionalization of graphene or polymer by ion implantation significantly increases its sensory properties [9,10]. The gas sensor morphology also significantly affects the sensing performance [11] and the combination of multi-energy ion implantation with ion microbeam provides the possibility to the preparation of spatially well-defined structures in both vertical and horizontal directions. The vertical structure also makes it possible to control the trans-conductance and sensitivity of the implanted layer by applying an external field. The aim of the present work is to give a further contribution to the understanding in the field of energetic ion beam preparation of spatially well-defined 3D metal-dielectric structures with potential applications in field-effect sensors and micro-electronic devices. The four different dielectric matrixes (PI, PET, PLLA, and GO) were implanted by Au ions with three different energies to create three independent layers of Au/insulator composite in polymer or graphene oxide and also for study of these prepared structures.

2 Experimental

Modified Hummers method was used for GO preparation as it is described in [12]. GO film was prepared by filtration on a polycarbonate membrane. Density of the prepared GO films, 1.36 $g\cdot cm^{-2}$, was determined by weighing a precisely cut-out part of the GO film. The 50-μm-thick foils of PI – Kapton, PLLA and PET – Mylar were supplied by Goodfellow, Ltd [13,14]. The prepared GO and polymer foils were implanted by the Au ions gradually with three different energies in the ascending or descending order. By this way two sets of samples were prepared, the first one irradiated by Au ions with energies 3.2, 1.6 and 0.8 MeV and the second one with energies 0.8, 1.6 and 3.2 MeV. The implantation was done in the implantation chamber installed at the 3 MV Tandetron MC 4130 accelerator by the ion current densities in the range 8–15 $nA.cm^{-2}$ and to the ion fluence 3.75×10^{14} cm^{-2}. Thus, each foil was implanted with a total Au-ion fluence 1.125×10^{15} cm^{-2} at various depths. Rutherford Back-Scattering spectrometry (RBS) and Elastic Recoil Detection Analysis (ERDA) were employed for the determination of GO and polymer composition and elemental depth profiling before and after the Au-ion irradiation. RBS spectra were collected using a beam of 2.0 MeV He^+ ions. An Ultra-Ortec PIPS detector recorded He^+ ions backscattered at a laboratory scattering angle of 170° in Cornell geometry. To reduce effects of the sample degradation during the RBS analysis, several particular spectra were measured on different beam spots and the final spectrum was obtained by summing the individual spectra. The ERDA spectra were measured using 2.0 MeV He^+ ions with the primary beam coming at an angle of 75° with respect to the foil surface normal and with hydrogen atoms recoiled at a scattering angle of 30° registered with the PIPS detector covered by a 12 μm Mylar foil. The typical ion current used during the RBS/ERDA analysis was 5 nA. SIMNRA code [15] was used for RBS spectra evaluation and O, H and Au elemental depth profiling. Energy-Dispersive X-ray Spectroscopy (EDS) was used to get deeper information about the GO and polymer foils elemental modification caused by the Au-ion irradiation. The EDS spectra were measured with the Scanning Electron Microscope (SEM) with a field emission gun electron source (TescanLyra dual beam microscope). For measurements, the samples were pasted onto carbon conductive tape and analysed using a 10 kV electrons. The surface morphology was determined with an Atomic Force Microscope (AFM, Dimension ICON, Bruker Corp.) in the quantitative nanoscale mechanical (QNM) mode on air. SCANASYST-AIR silicon tip on nitride lever with a spring constant of 0.4 $N \cdot m^{-1}$ was used. NanoScope Analysis software was applied for data processing. The characteristics of the electric properties of pristine and irradiated samples were studied by the standard 2-point method utilizing the Keithley 6221 Ultra-sensitive Current Source and Keithley 6517B Electrometer/High Resistance Meter. For the sheet resistance measurement two Au contacts, 50 nm thick and 10 mm long, with relative distance 1 mm were sputtered on the surface of GO and polymer foils.

3 Results and Discussion

Electronic energy stopping power of implanted ions leads to atom ionization, electron excitation, bond disruption and subsequent creation of free radicals [7]. Contrary, the nuclear energy stopping power leads to atom knocking and vacancies production [16]. The electronic stopping in polymers containing heteroatoms leads also to the gaseous low mass fragments release and subsequent to dehydrogenation [17]. We made the initial estimation of the Au-ion energy losses and ranges in GO, PI, PET and PLLA foils by the SRIM code [16], where the calculation was performed using the full cascade MC simulation. The GO initial composition was based on the real sample analysis. The composition and properties of polymers were obtained from the SRIM compound dictionary and from [13,14]. Table 1 displays ratios of nuclear to electronic stopping power (S_n/S_e) and ranges of the Au-ions for all energies. It is clear that the nuclear stopping plays a main role in this case and we can expect a massive bond rupturing, low-mass fragments release deoxygenation and dehydrogenation of the implanted layer. The threshold from which carbonisation occurs in polymers is 150–200 eV/nm [18] and we can predict from SRIM simulation significant increasing of carbon concentration that should lead to the conductivity enhancement. On the Fig. 1, one can see the chosen distributions of the SRIM calculated Au depth profiles of 3.2, 1.6 and 0.8 MeV Au ions in GO and PLLA together with the nuclear and electronic stopping power depth profiles for illustration. The summary of the projected ranges Rp is presented in Table 1. The mean projected range of 3.2 MeV Au ions is ~ 1000 nm, 1.6 MeV Au-ions reach the mean depth higher than 500 nm and 0.8 MeV Au ions most often stop before reaching of 300 nm and is evident that Au would create three well distinguishable layers. The influence of the nuclear stopping power significantly grows with the increasing depth and the Au-ion descending energy.

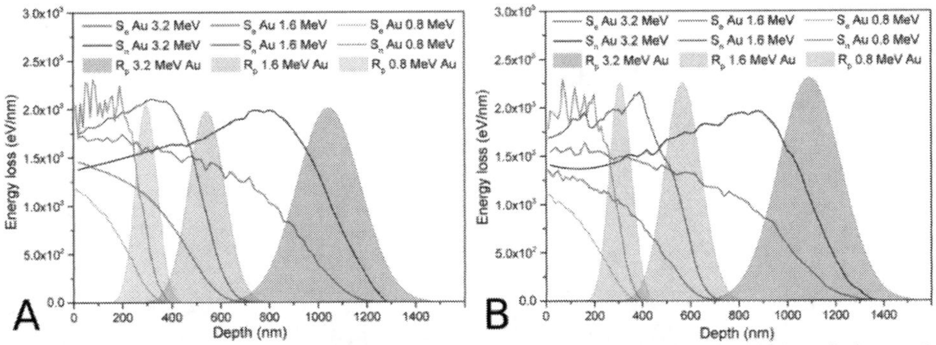

Figure 1. The SRIM calculated Au-depth distribution, nuclear stopping powers (S_n) and electronic (S_e) stopping powers of Au ions with energies 3.2, 1.6 and 0.8 MeV in GO (A) and PLLA (B).

Table 1. The ratios of nuclear and electronic stopping of Au ions and their projected ranges (Rp) with energies 3.2, 1.6 and 0.8 MeV in GO, PLLA, PI and PET matrix.

	Au 3.2 MeV S_n/S_e	Au 1.6 MeV S_n/S_e	Au 0.8 MeV S_n/S_e	Rp (nm) 0.8/1.6/3.2 MeV
GO	1.33	1.85	2.52	297/542/1044
PI	1.32	1.82	2.46	276/502/968
PET	1.42	1.96	2.62	280/514/999
PLLA	1.38	1.95	2.65	309/568/1093

The elemental composition of the pristine and irradiated samples was determined by RBS and ERDA (Fig 2, Table 2). The elemental concentration of C, O and H is presented in the Table 2 and the chosen spectra for GO and PLLA shows the Fig 2 together with the Au-depth profiles. The oxygen release is observable in the implanted GO that is more pronounced for the descending energies (3.2-1.6-0.8 MeV), (Fig. 2A). The PI exhibits the similar behaviour as the GO. The pronounced oxygen release is observable in PLLA only after Au implantation with descending energies (Fig. 2B). The Au implantation with ascending energies (0.8-1.6-3.2 MeV) leads to the O release in PLLA just in the subsurface layer. The behaviour of PET reports such progress as of PLLA. The similar behaviour of the GO and PI versus to the PET and PLLA is connected with the chemical structures. The GO and PI have more aromatic rings in the structure compare to the PET and PLLA. The hydrogen concentration decreases in polymers after Au implantation. This decline is more pronounced after implantation with ascending energies and the most pronounced in PLLA. Contrary, the enrichment of H concentration can be seen in GO, which may be due hydrogenation of the irradiated surface from ambient atmosphere, as observed earlier [19]. The breaking of the volatile compounds and its release leads to the carbonisation, to the accompanying formation of conjugated bonds and conductivity enhancement [18]. The Au depth profiles in GO and PLLA are presented in Figs 2C, D. The position of the Au depth profile in PLLA and in GO implanted with ascending energies well agrees with simulated one by SRIM. The merging of Au profiles in GO implanted with descending energies is connected with the GO high surface roughness and its influence on the RBS analysis. The Au depth profile slight shift to the depth appears in GO, PLLA and PET after implantation with descending energies, which can be connected with sputtering of samples surface layers by 0.8 MeV Au in the last of implantation. It is evident, that the Au creates three well distinguishable independent layers usable in further 3D sensory structures preparation.

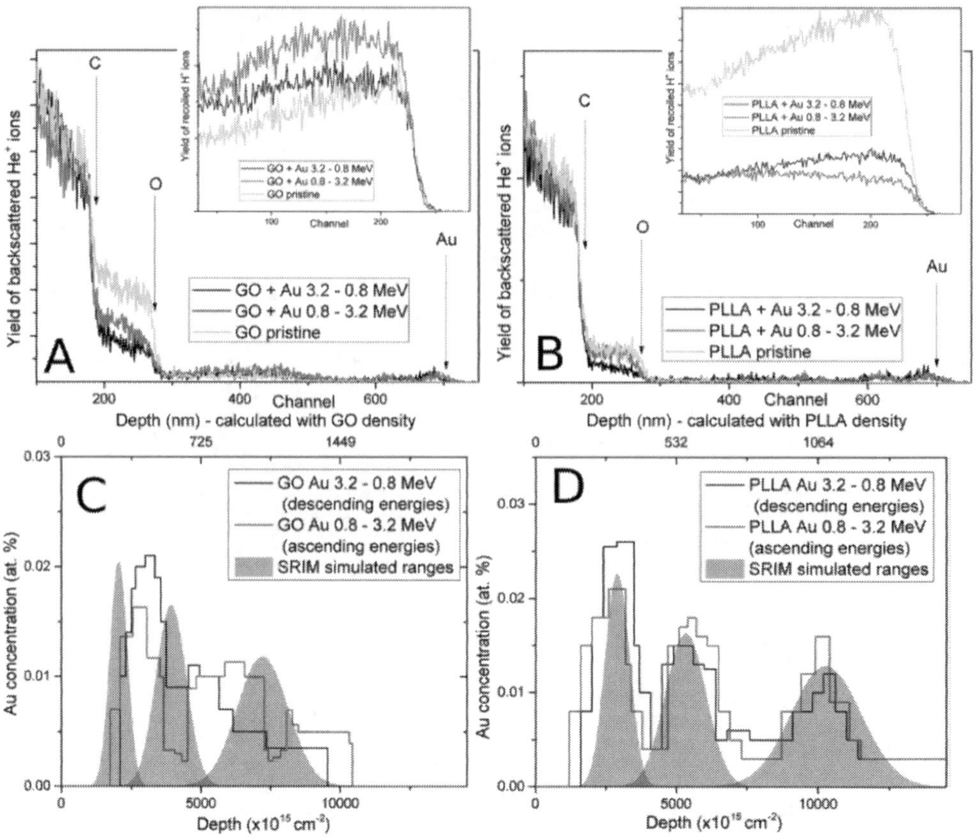

Figure 2. The RBS and ERDA spectra as insets are presented in A, B and detailed Au-depth profiles as-evaluated from RBS with SRIM simulated ranges are presented in(C, D) (inset) of Au ions with energies 3.2, 1.6 and 0.8 MeV in GO and PLLA.

Table 2. The C, O and H concentrations and their ratios on surface (**SU**) and in the depth of ~250 nm (**DE**) in pristine and Au irradiated GO and polymers obtained by RBS and ERDA.

RBS/ ERDA		Pristine					0.8->3.2 MeV ascending energies					3.2->0.8 MeV descending energies				
		C	O	H	C:O	C:H	C	O	H	C:O	C:H	C	O	H	C:O	C:H
GO	**SU**	65.9	21.6	9.8	3.1	6.7	77.6	7.1	11.8	10.9	6.6	78.8	7.1	10.2	11.1	7.7
	DE						71.3	10.4	14.9	6.9	4.8	76.6	7.2	12.5	10.6	6.1
PET	**SU**	53.4	9.3	37.3	5.7	1.4	79.0	4.3	16.6	18.4	4.8	74.0	2.8	23.2	26.4	3.2
	DE						73.3	9.6	17.1	7.6	4.3	73.2	2.8	24.0	26.1	3.1
PLLA	**SU**	53.3	5.3	41.4	10.1	1.3	87.2	3.3	9.6	26.4	9.1	79.5	3.4	17.1	23.4	4.6
	DE						82.2	5.8	12.0	14.1	6.9	79.2	3.3	17.5	24.0	4.6
PI	**SU**	56.4	12.8	25.6	4.4	2.2	71.6	6.9	15.9	10.4	4.5	74.6	3.1	16.8	24.0	4.4
	DE						71.8	7.0	15.6	10.3	4.6	74.9	3.2	16.2	23.4	4.6

The Energy Dispersive Spectroscopy (EDS) provides a qualitative evaluation of the elemental composition changes between the non-irradiated and irradiated parts of the substrates (see Table 3). The accessible information depth of EDS using 10 keV electrons in our samples is approximately 1 μm, which is comparable with that of RBS. RBS is able to provide information from individual depths, while the EDS tells just about the integral amount of elements in the whole analysed depth. The oxygen release and associated carbonization are observed in the EDS after the Au-ion implantation, which is more pronounced for the samples implanted with descending energies. This finding is in agreement with above described RBS results.

Table 3. The C, O concentrations and their ratios in pristine and Au irradiated GO and polymers obtained by EDS.

EDS	Pristine			0.8->3.2 MeV ascending energies			3.2->0.8 MeV descending energies		
	C	O	C:O	C	O	C:O	C	O	C:O
GO	65.6	34.1	1.9	84.1	15.3	5.5	86.1	13.2	6.5
PET	71.9	28.1	2.6	85.3	14.7	5.8	92.9	7.1	13.1
PLLA	61.9	38.0	1.6	88.7	11.2	7.9	95.2	4.6	20.7
PI	82.0	18.0	4.6	85.1	12.4	6.9	90.0	6.3	14.3

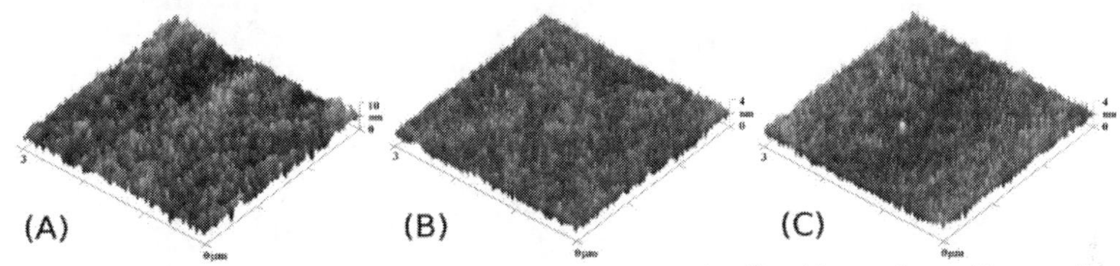

Figure 3. The 3D AFM image of surface of pristine PI (A), PI irradiated by Au ions with ascending energies (B) and PI irradiated by Au ions with ascending energies (C).

The surface morphology changes, as studied by AFM, are introduced in Table 3, where the roughness parameters are presented for all samples and in the Fig 3, where the 3D surface of chosen PI is shown for ilustration. The pristine GO surface exhibits the morphology with obvious irregularities and significant roughness, which increases even more after implantation. The roughness increasing is more pronounced for GO implanted by Au ions with descending energies, which can be connected with the GO surface sputtering with the last Au-ion implantation with lowest energy 0.8 MeV, as also precluded from RBS. As a result of the ion implantation, the roughness also dramatically increase on the surface of PET and the only mild roughness increasing has been observed on the PLLA surface. Rather different situation is observed for the PI, where the Au-ion implantation leads to the polymer rougness decreasing. The ion irradiation leads to the surface roughness change of polymes based on their chemical structure. PI as a polymer with complex monomer shows the surface smoothening under the ion irradiation. Contrary, in the polymers with simpler structure (PET, PLLA) can the chain scission leads to surface roughness increasing [20].

Table 4. The surface roughness arithmetic average height parameters (R_a) and the mean roughness parameters (RMS) of pristine and modified samples as obtained by AFM.

AFM	Pristine		0.8->3.2 MeV ascending		3.2->0.8 MeV descending	
	R_a	RMS	R_a	RMS	R_a	RMS
GO	49.2	61.6	65.6	79.8	99.6	126
PET	0.46	0.62	0.64	0.79	0.67	0.88
PLLA	6.5	7.3	8.3	10.4	6.9	8.5
PI	2.33	3.45	0.32	0.43	0.53	0.72

The Au-ion implanted samples sheet resistances, displayed in Figure 4, were measured to get information about the changes in electrical properties. The conductivity of all samples increases after the Au implantation in comparison with that of pristine. In the PET, GO and PLLA the conductivity increase is more pronounced for samples implanted with Au-ions ascending energies. The changes in the electrical conductivity after implantation can be attributed to the carbonization of thin subsurface layer, as proved by ERDA/RBS measurements and predicted by SRIM, and to the subsequently growth of the carbon clusters, which form a conjugated network. A conductive network with many cross-linked carbon connections then facilitates charge carrier transfer in the modified surface. The conductivity enhancement in polymers after the change of energies sequence is related to the higher concentration of carbon on the surface and the hydrogen concentration decrease, and is more pronounced in PPLA and PET with ascending energies and in PI with descending energies. Contrary, the conductivity of the GO after implantation with ascending energies is better in spite of lower carbon concentrationin the surface. This trend can be connected with GO porousity and islet structure, which can caused higher influence of the deeper layers, where the carbon concentration is higher.

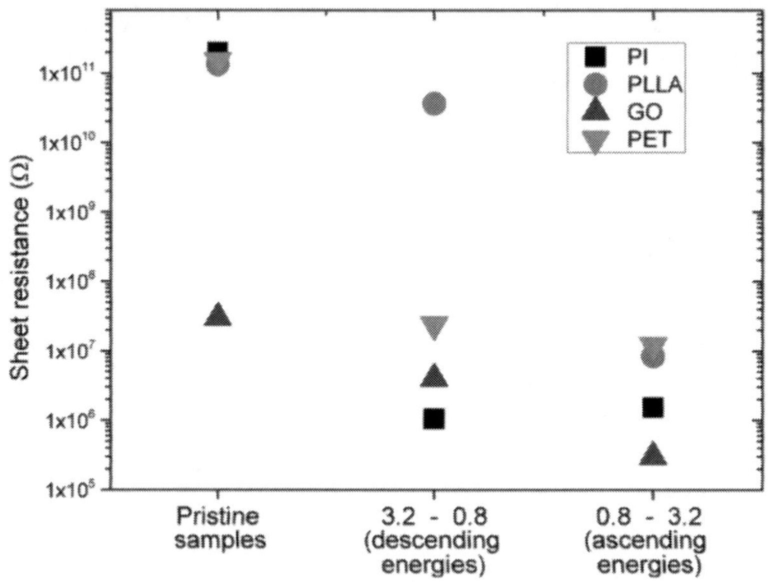

Figure 4. The electric resistance of GO, PLLA, PI and PET before and after Au ion irradiation.

4 Conclusion

In this work, the 3D Au-dielectric structures were synthetized by ion implantation in PI, PET, PLLA and GO with potential applications in the field-effect micro-sensory and micro-electronic application. We proved, that the conductive multilayer system with Au-doped and un-doped parts can be created in dielectric samples by ion implantation. The best quality of Au distribution was found in PLLA with three well separated layers. We can say that the Au-ion implantation with descending energies has more significant influence on the oxygen composition, while the ascending energies order influence little bit more the hydrogen concentration in polymers. The electric conductivity enhancement in polymers is connected with carbonization of the surface layer, caused be decreasing of both, oxygen as well as hydrogen concentration. It seems that Au implantation starting with higher energy causes more damage. The descending Au-ion energies being implanted mean, that as a first the deeper layer is irradiated, and then the next sub-sequent Au-ion implantations are provided into the modified layer and may cause more damage. Opposite situation is in case of the ascending energy implantation, where in the first step the surface is significantly modified by the Au-ions with the lowest energy of 0.8 MeV and this partially carbonized layer with gold particles inside can than play a function of sample protection.

The research has been carried out at the CANAM (Centre of Accelerators and Nuclear Analytical Methods) infrastructure LM 2015056 and promoted by the Czech Science Foundation (GACR No. 19-02482S) and University of J.E. Purkyne project UJEP-SGS-2021-53-005-2. This publication was supported by OP RDE, MEYS, Czech Republic under the project CANAM OP, CZ.02.1.01/0.0/0.0/16_013/0001812.

References

1. J. Boudaden, M. Steinmaß, H.E. Endres, A. Drost, I. Eisele, Ch. Kutter, P. Müller-Buschbaum, Sensors **18**, 1516, (2018)
2. D. Zhang, J. Tong, B. Xia, Q. Xue, Sensors and Actuators B **203**, 263–270, (2014)
3. T. Wang, D. Huang, Z. Yang, S. Xu, G. He, X. Li, N. Hu, G. Yin, D. He, L. Zhang, Nano-Micro Lett. **8**, 95–119, (2016)
4. N.M. Nurazzi, N. Abdullah, S.Z.N. Demon, N.A. Halim, A.F.M. Azmi, V.F. Knight, I.S. Mohamad, Nanotechnology reviews **10**, 330-369, (2021)
5. F.A.A. Nugroho, I. Darmadi, L. Cusinato, A. Susarrey-Arce, H. Schreuders, L.J. Bannenberg, A.B. da Silva Fanta, S. Kadkhodazadeh, J.B. Wagner, T.J. Antosiewicz, Anders Hellman, V.P. Zhdanov, B. Dam, Ch. Langhammer, Nature materials 18, 489-495, (2019)
6. V. Resta, G. Quarta, I. Farella, L. Maruccio, A. Cola, L. Calcagnile, Nucl Instrum Meth B **331**, 168-171, (2014)
7. V. Popok ,Rev Adv Mater Sci.30, 1-26, (2012)
8. M. Döbeli , T.J. Jones , A. Lee , R.P. Livi,T.A. Tombrello, Radiation Effects and Defects in Solids **118**, 325-339, (1991)
9. R. Kumar, D.K. Avasthi, A Kaur, Sensors and Actuators B **242**, 461-468, (2017)
10. C.R. Minitha, L.R. Nivedita, K. Asokan, R.T.R. Kumar, Nucl Instrum Meth B **450**, 257-261, (2019)
11. A. Mirzaei, J.H. Lee, S.M. Majhi, M. Weber, M. Bechelany, H.W. Kim, S.S. Kim, J. Appl. Phys.**126**, 241102, (2019)
12. O. Jankovský, P. Šimek, J. Luxa, D. Sedmidubský, I. Tomandl, A. Macková, R. Mikšová, P. Malinský, M. Pumera, Z. Sofer, Chem Plus Chem **80**, 1399–1407, (2015)
13. http://www.goodfellow.com
14. W. Martienssen, H. Warlimont, Springer Handbook of Condensed Matter and Materials Data, Springer-Verlag, Berlin, Heidelberg, 2005, 978-3-540-30437-1.
15. M. Mayer, SIMNRA version 6.06, Max-Planck-Institut fur Plasmaphysik Garching, Germany, 2006 Available at: http://www.rzg.mpg.de/~mam/
16. J.F. Ziegler, Nucl Instrum Meth B. **219–220**, 1027-1036, (2004)
17. D.V. Sviridov, Russ Chem Rev. **71**, 315–327, (2002)
18. V.N. Popok, I.I. Zarko, R.I. Khaibullin, A.L. Stepanov, V. Hnatowicz, A. Mackova, S.V. Prasalovich, Appl. Phys. A **78**, 1067–1072, (2004)

19. P. Malinský, M. Cutroneo, A. Macková, V. Hnatowicz, M. Florianová, M. Boháčová, D. Bouša, Z. Sofer, Surface and Coatings Technology **342**, 220–225, (2018)
20. R. Mikšová, A. Macková, M. Cutroneo, P. Slepička, J. Matoušek, Surface and Interface analysis **48**, 566-569, (2016)

Extension of the BIANCA biophysical model up to Fe-ions and applications for space radiation research

Ricardo Ramos[1*], *Alessia* Embriaco[1,2], *Mario* Carante[1,3], *Alfredo* Ferrari[4,5], *Paola* Sala[6], *Valerio* Vercesi[1] *and Francesca* Ballarini[1,3]

[1]INFN (Italian National Institute for Nuclear Physics), Sezione di Pavia, via Bassi 6, I-27100 Pavia, Italy

[2]ENEA, Istituto Nazionale di Metrologia delle Radiazioni Ionizzanti, Roma Italy

[3]University of Pavia, Physics Department, via Bassi 6, I-27100 Pavia, Italy

[4]University Hospital Heidelberg, Germany

[5]Gangneung-Wonju National University, Gangneung, Korea

[6]INFN (Italian National Institute for Nuclear Physics), Sezione di Milano, via Celoria 16, I-20133, Milano, Italy

Abstract. BIANCA (BIophysical ANalysis of Cell death and chromosome Aberrations) is a biophysical model, implemented as a Monte Carlo code, which simulates the induction of chromosome aberrations and cell death by different monochromatic ion beams (i.e., different ion types and energy values), as well as photons. In previous works, the model predictions for cell survival and lymphocyte dicentrics along therapeutic-like ion beams have been successfully benchmarked against experimental data. With the aim of evaluating the biological damage induced by Galactic Cosmic Rays (GCR), in this study BIANCA was extended up to Fe-ions. A radiobiological database describing human skin fibroblast cell survival and lymphocyte dicentrics as a function of ion type ($1 \leq Z \leq 26$) and energy, as well as dose, was constructed. Afterwards, interfacing BIANCA with the FLUKA Monte Carlo transport code, a feasibility study was performed to calculate the Relative Biological Effectiveness (RBE) of Galactic Cosmic Rays, both for dicentrics and for cell survival; the results were discussed with respect to available space radiation protection limits. Following this work, BIANCA can now provide RBE predictions of cell killing, which can be related to deterministic effects, and lymphocyte dicentrics, more related to stochastic effects, for space radiation exposure.

*Corresponding author: ricardo.ramos@pv.infn.it

1 Introduction

Recently, space research has gained a renewed interest, also due to the fact that NASA is planning to go back to the Moon with a human mission and that the Moon might represent an intermediate step towards Mars. Human space missions are subject to many risks, including those due to space radiation exposure [1]. Galactic Cosmic Rays (GCR), which consist of about 87% protons, 12% He ions and 1% heavier ions in fluence delivered at a relatively low dose-rate (about 1 mSv/day in free space), can represent a serious hazard for the crewmembers of long-term missions, especially outside the Geomagnetic field. In particular, heavy ions like Fe are known to be very effective at inducing DNA damage [e.g. 2], which in turn may lead to higher level damage involving chromosomes, cells and tissues/organs [e.g. 3-5]. The exposure scenario is further complicated by the so-called "Solar Particle Events" (SPEs), which are occasional injections of high fluxes (up to more than 10^{10} particles cm^{-2} in few hours) of charged particles coming from the Sun, mainly protons with energy lower than a few hundred MeV.

Many works are available in the literature on the calculation of astronauts' doses [e.g. 6]. In the past, the FLUKA Monte Carlo transport code [7-10] has been coupled with simulated anthropomorphic phantoms inserted into an aluminium shell of variable thickness; according to those calculations [10], the GCR effective dose during a 2-years mission in deep space (typical duration of a Mars mission) at solar minimum would be about 1 Sv. Furthermore, in case of a SPE similar to the August 1972 one, a 10 g/cm^2 Al storm shelter would allow to respect the equivalent dose limits for skin, eye and blood forming organs reported in the NCRP report n. 132 for 30-days missions in Low Earth Orbit [11]. Most experimental data on space radiation biological damage come from experiments performed at facilities like that available in Brookhaven, where cells and small animals can be irradiated by different heavy ion beams up to Fe. Information on the effects of space radiation on humans also comes from the observation of chromosome aberrations in peripheral blood lymphocytes (PBL) taken from astronauts' blood samples [e.g. 6,12,13]. In these measurements, pre-flight calibration curves obtained by *in vitro* gamma-ray irradiation are used to convert the post-flight aberration yields into equivalent doses. In general, such biodosimetry estimates lie within the range expected from physical dosimetry. In this work the BIANCA (BIophysical ANalysis of Cell death and chromosome Aberrations) biophysical model, which simulates the induction of chromosome aberrations and cell death by photons and different (monochromatic) ion beams, was extended up to Fe ions, also considering that a new source capable of accelerating Fe ions will be available soon at CNAO (Pavia, Italy), in the framework of the INSpIRIT project. Following interface to FLUKA, this allowed predicting RBE values (both for cell death and for chromosome aberrations) following GCR exposure, thus providing the bases for future calculations on space radiation health risk.

2 Materials and Methods

BIANCA is a biophysical model, implemented as a Monte Carlo code, which simulates the induction of cell death and chromosome aberrations by different monochromatic ion beams, as well as photons. Although detailed descriptions of the model can be found in previous works [e.g. 14-17], it is worth mentioning that BIANCA is based on the following main assumptions: i) ionizing radiation can induce DNA "Critical Lesions" (CLs), where a CL is defined as a lesion that interrupts the chromatin fibre producing two (main) independent fragments; ii) distance-dependent incorrect rejoining of such fragments, or fragment un-rejoining, produces chromosomal aberrations; iii) certain aberration types (dicentrics, rings and large deletions, where "large" means visible when chromatin is condensed) lead to cell death. The CL yield and the fragment un-rejoining fraction (f) are the only two adjustable parameters. In previous works [15,18,19], an approach has been developed to predict cell survival following ion irradiation of an arbitrary cell line for which the photon dose-response is known, starting from a reference cell line; V79 cells have been chosen as a reference.

In the present work, based on the parameter values derived for V79 cells and the formalism described in [15], survival curves for human skin fibroblasts (HSF) exposed to different monoenergetic ion beams (protons, He-ions, C-ions, Ne-ions and Fe-ions) were predicted. These cells were chosen because they represent the radiation response of the skin, which is one of the organs that have to be considered for radiation protection. Briefly, to simulate HSF survival after irradiation with a given ion type of given energy (and thus LET), we derived the values of the two model parameters (CL and f) by rescaling the parameters previously used to reproduce the survival of V79 cells, which have been chosen as a reference; further details can be found in [15]. Afterwards, each of the (simulated) HSF survival curves was fit by the well-known linear-quadratic equation:

$$S(D) = \exp(-\alpha D - \beta D^2) \qquad (1)$$

where S(D) is the fraction of surviving cells after receiving an absorbed dose D, and α and β are fitting coefficients. These coefficients were then stored in a table constituting a radiobiological database that allows predicting HSF cell survival as a function of ion type and energy.

Concerning chromosome aberrations, we developed an analogous database that allows simulating dicentric dose-response curves in human lymphocytes. Specifically, the CL parameter has been adjusted to reproduce experimental dose-responses taken from the literature for photons and for different monochromatic ion beams (protons, He-ions, C-ions, N-ions, O-ions, Ne-ions and Fe-ions); afterwards, the LET-dependence of the CL yield has been fit for each ion type, thus allowing us to predict dose-response curves for, in principle, any LET value. Finally, each of these (simulated) curves has been fit by the following linear-quadratic function, which is considered as a good description for dicentric induction unless the dose is too high (about 10 Gy for photons, a few Gy for high-LET radiation):

$$Y(D) = \alpha D + \beta D^2 \qquad (2)$$

In eq. (2), Y(D) is the mean number of dicentrics per cell after an absorbed dose D, and α and β are fitting coefficients. Such coefficients, together with those describing the photon dose-response, have been stored in a table that describes lymphocyte dicentric dose-response for different monochromatic ion beams, as well as photons.

Finally, the radiobiological database for lymphocyte dicentrics and that for HSF cell survival were read by FLUKA, exploiting a pre-existing interface between BIANCA and FLUKA. Specifically, whenever according to FLUKA a certain amount of energy (and thus a certain dose, D_i) was deposited in a target voxel by a given particle type of given energy, FLUKA read from the tables the corresponding linear and quadratic coefficients (α_i and β_i), and used them to calculate the average coefficients (α and β) describing the fraction of surviving cells or the yield of dicentrics by the mixed field in that voxel:

$$\alpha = \frac{\sum_{i=1}^{n} \alpha_i D_i}{\sum_{i=1}^{n} D_i} \quad (3)$$

$$\sqrt{\beta} = \frac{\sum_{i=1}^{n} \sqrt{\beta_i} D_i}{\sum_{i=1}^{n} D_i} \quad (4)$$

Afterwards, the RBE in each voxel was calculated as D_X/D, where D is the total absorbed dose in the voxel and D_X for cell survival and for dicentric induction was calculated as follows, respectively:

$$D_X = \frac{\left[-\alpha_X + \sqrt{\alpha_X^2 + 4\beta_X \ln S}\right]}{2\beta_X} \quad (5)$$

$$D_X = \frac{\left[-\alpha_X + \sqrt{\alpha_X^2 + 4\beta_X Y}\right]}{2\beta_X} \quad (6)$$

More details can be found in [18-21]. This allowed performing a feasibility study to calculate the Relative Biological Effectiveness (RBE) of Galactic Cosmic Rays (GCR), both for dicentric induction and for cell survival.

3 Results and discussion

In order to obtain the α and β coefficients for HSF cell survival, we simulated 14 survival curves for proton beams (with LET in the range 2.5-35 keV/μm), 20 curves for He-ion beams (LET range: 5-200 keV/μm), 30 curves for C-ion beams (LET range: 10-500 keV/μm), 25 curves for Ne-ion beams (LET range: 20-500 keV/μm), and 26 curves for Fe-ion beams (LET range: 75-900 keV/μm). Fig.1 shows the α and β values obtained for HSF cell survival as a function of LET for protons, He-ions, C-ions, Ne-ions and Fe-ions. As expected, unless for very high LET values where the so-called overkilling dominates, the values of α increase with LET up to 100-200 keV/μm for all considered ions. Furthermore, lighter ions tend to show higher α values with respect to heavier ions having the same LET; this reflects the fact that lighter ions, having lower energy, produce lower-energy secondary electrons and thus are characterized by a denser track structure, which induces DNA damages that are more clustered and thus more difficult to be repaired by the cell. The interpretation of the behaviour of the β coefficient is less straightforward: while the very low values found at high LET confirm that at high LET the survival curve is basically linear, the increase shown by some ion types at lower LET values is more difficult to explain; however, it is worth mentioning that this increase is much less pronounced than the increase shown by coefficient α.

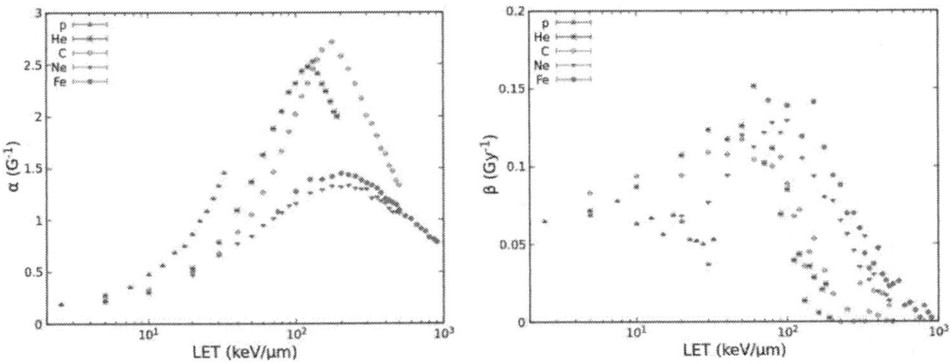

Fig. 1. Values of the coefficients α (left panel) and β (right panel) describing HSF cell survival as a function of LET for the different ion beams considered in this work.

Concerning the induction of dicentric chromosomes in human lymphocytes, we simulated 14 curves for protons (LET range: 2.5-35 keV/μm), 12 curves for He-ions (LET range: 5-110 keV/μm), and 16 curves for heavy ions between carbon and iron (LET range: 5-150 keV/μm). Fig.2 shows the values of the α and β coefficients obtained by fitting each of these curve with eq. (2). Importantly, while the β coefficients for dicentrics are in the same order than those for cell survival, the dicentric α coefficients are much smaller than those for cell survival. This implies that dicentrics have a lower α/β ratio, which is consistent with the higher degree of curvature shown by lymphocyte dicentric dose-responses with respect to survival curves.

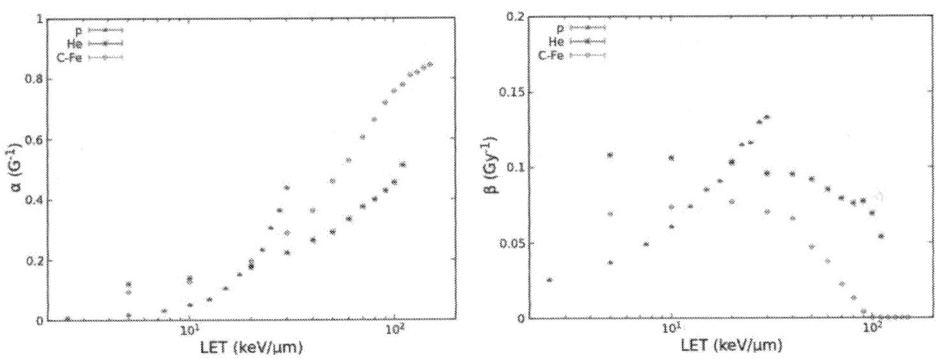

Fig. 2. Values of the coefficients α (left panel) and β (right panel) for lymphocyte dicentrics as a function of LET for the different ion beams considered in this work.

As an exercise in the framework of biological damage evaluation following space missions, the absorbed dose in a spherical water phantom (radius: 15 cm) included in a spherical, isotropic source (radius: 32 cm) was calculated by the FLUKA code. For these simulations, the GCR spectrum at solar minimum conditions embedded in FLUKA was considered. The RBE for lymphocyte dicentrics and that for cell death were calculated by reading the two radiobiological databases described in the Methods. The spherical phantom was irradiated by a total absorbed dose of 0.5 Gy, which is in the order of the dose that should be received by an astronaut during a Mars mission [e.g. 22]. Table 1 shows the RBE values and the RBE-weighted doses calculated for lymphocyte dicentrics (columns 2 and 3) and for HSF cell death (columns 4 and 5).

Table 1. RBE values and RBE-weighted doses for lymphocyte dicentrics (RBE_{dic} and $Dose*RBE_{dic}$) and for HSF cell death (RBE_{surv} and $Dose*RBE_{surv}$) induced by GCR at solar minimum for a total absorbed dose of 0.5 Gy.

	RBE_{dic}	$Dose*RBE_{dic}$ ($Gy*RBE_{dic}$)	RBE_{surv}	$Dose*RBE_{surv}$ ($Gy*RBE_{surv}$)
GCR	2.10	1.05	1.90	0.95

This exercise showed that, at least for the considered dose, the RBE for lymphocyte dicentrics tends to be higher than that for skin fibroblast cell death. This may be explained by considering that the photon dose-response for lymphocyte dicentrics is characterized by a much lower α/β ratio with respect to (fibroblast) cell survival curves, thus implying that the response for lymphocyte dicentrics has a more pronounced curvature that leads to a higher RBE at low and intermediate doses. Based on this interpretation, the difference between the RBE for dicentrics and that for cell death should increase by decreasing the dose.

A comparison with the astronauts' 10-year career limits for stochastic effects recommended by NCRP [11], which are age- and sex-dependent, shows that the calculated RBE-weighted dose for dicentrics (1.05 GyRBE) is higher than the limit for 35-years-old male astronauts (which is 1.0) and that for 45-years-old female astronauts (which is 0.9). Concerning the limits for deterministic effects, the RBE-weighted dose for skin fibroblast cell survival found in this work (0.95 GyRBE) is even lower than the 30-day-mission limit recommended by NCRP for skin, which is 1.5 GyEq. On the contrary the value found for lymphocyte dicentrics is higher than the limits for blood forming organs, which are 0.25 GyEq for a 30-days mission and 0.5 GyEq for a 1-year mission. Of course these comparisons have to be taken with caution for several reasons, including the following: 1) the limits for stochastic effects are effective doses expressed in Sv, whereas the numbers calculated in this work are RBE-weighted doses expressed in GyRBE; 2) the phantom used in this work was a water sphere, whereas in the future we plan to repeat these calculations for an anthropomorphic phantom; 3) this work did not consider any shielding, which will be included in future studies and is expected to lead to decreased doses; 4) since lymphocyte chromosome aberrations are better correlated with cancer incidence [e.g. 23,24], a comparison of the results found in this work for lymphocyte dicentrics with the blood forming organ (BFO) limits for deterministic effects is not straightforward.

4 Conclusions

The BIANCA biophysical model was extended to heavy ions up to Fe, in view of studies on space radiation effects. Specifically, two radiobiological databases were constructed for these ions: the first one allows predicting cell death for human skin fibroblasts, whereas the second one describes lymphocyte dicentrics, which are considered as indicators of stochastic damage including cancer. As a pilot study, BIANCA was interfaced with the FLUKA code and RBE values (both for lymphocyte dicentrics and for HSF cell death) were calculated for a water sphere irradiated by a GCR spectrum; the results were compared with astronauts' dose limits recommended by NCRP. This work thus allowed developing a tool that, when interfaced to a radiation transport code capable of handling the various space radiation components, can provide RBE predictions of cell death and lymphocyte dicentrics not only for ion therapy, but also for space radiation exposure. As a future development, we plan to include the effects of shielding and, possibly, to adopt a more realistic anthropomorphic phantom.

This work was supported by INFN (project MC-INFN/FLUKA), the University and Research Ministry ("MIUR Dipartimenti di Eccellenza 2018-2022 project F11I18000680001"), and the INSpIRIT project by Regione Lombardia, ID 1161908, co-funded on resources POR FESR 2014-2020-Asse 1.

References

1. Marco Durante and F Cucinotta, Physical basis of radiation protection in space travel. Reviews of modern physics 83, 1245-1281, 2011
2. A.Campa, D. Alloni, F. Antonelli, F. Ballarini, M. Belli, V. Dini, G. Esposito, A. Facoetti, W. Friedland, Y. Furusawa, M. Liotta, A. Ottolenghi, H.G. Paretzke, G. Simone, E. Sorrentino, M. A. Tabocchini (2009), DNA fragmentation induced in human fibroblasts by ^{56}Fe ions: experimental data and MC simulations. Radiat. Res. 171, 438-445
3. A.Ottolenghi, F. Ballarini, M. Merzagora (1999), Modelling radiation induced biological lesions: from initial energy depositions to chromosome aberrations. Radiat. Environ. Biophys. 38, 1-13
4. A.Ottolenghi, F. Ballarini, M. Biaggi (2001), Modelling chromosomal aberration induction by ionising radiation: the influence of interphase chromosome architecture. Adv. Space Res. 27, 369-382.
5. F. Ballarini, M. Biaggi, A. Ottolenghi (2002), Nuclear architecture and radiation-induced chromosome aberrations: models and simulations. Radiat. Prot. Dosim. 99, 175-182.
6. F Cucinotta, M-H Kim, V Willingham, K George, Physical and biological organ dosimetry analysis for international space station astronauts. Radiat Res 170,170(1):127-38, 2008.
7. Ferrari A, P.R. Sala, A. Fassò and J. Ranft, FLUKA: a multi-particle transport code, CERN 2005-10 (2005), INFN/TC_05/11, SLAC-R-773
8. F. Ballarini, G. Battistoni, M. Campanella, M. Carboni, F. Cerutti, A. Empl, A. Fassò, A. Ferrari, E. Gadioli, M.V. Garzelli, M. Lantz, M. Liotta, A. Mairani, A. Mostacci, S. Muraro, A. Ottolenghi, M. Pelliccioni, L.S. Pinsky, J. Ranft, S. Roesler, P.R. Sala, D. Scannicchio, S. Trovati, R. Villari, T. Wilson, N. Zapp, V. Vlachoudis (2006), The FLUKA code: an overview. J. Phys. Conf. Series 41, 151
9. Bohlen TT, F. Cerutti, M.P.W. Chin, A. Fassò, A. Ferrari, P.G. Ortega, A. Mairani, P.R. Sala, G. Smirnov and V. Vlachoudis, The FLUKA Code: Developments and Challenges for High Energy and Medical Applications. Nuclear Data Sheets 120, 211-214 (2014)
10. F. Ballarini, G. Battistoni, F. Cerutti, A. Fassò, A. Ferrari, E. Gadioli, M.V. Garzelli, A. Mairani, A. Ottolenghi, H.G. Paretzke, V. Parini, M. Pelliccioni, L. Pinsky, P. R. Sala, D. Scannicchio, S. Trovati, M. Zankl (2006). GCR and SPE organ doses in deep space with different shielding: Monte Carlo simulations based on the FLUKA code coupled to anthropomorphic phantoms. Adv. Space Res. 37(9), 1791-1797

11. National Council on Radiation Protection and Measurements, Recommendations of Dose Limits for Low Earth Orbit, NCRP Report 132, NCRP, Bethesda, M. D., 2000.
12. Durante Marco, Riv. Nuovo Cimento, **25** (2002) 1.
13. F. Ballarini, A. Ottolenghi'(2005), A model of chromosome aberration induction: applications to space research. Radiat. Res. 164(4), 567-570.
14. Francesca Ballarini and Mario P Carante. Chromosome aberrations and cell death by ionizing radiation: evolution of a biophysical model. Radiation Physics and Chemistry, 128:18-25, 2016.
15. Mario Pietro Carante, Chiara Aimè, John James Tello Cajiao and Francesca Ballarini (2018), BIANCA, a biophysical model of cell survival and chromosome damage by protons, C-ions and He-ions at energies and doses used in hadrontherapy. Physics in Medicine and Biology 63/7, 075007.
16. John James Tello Cajiao, Mario Pietro Carante, Mario Antonio Bernal Rodriguez, and Francesca Ballarini. Proximity effects in chromosome aberration induction by low-LET ionizing radiation. DNA repair, 58:38-46, 2017.
17. John James Tello Cajiao, Mario Pietro Carante, Mario Antonio Bernal Rodriguez, and Francesca Ballarini. Proximity effects in chromosome aberration induction: Dependence on radiation quality, cell type and dose. DNA repair, 64:45-52, 2018.
18. M.P. Carante, G. Aricò, A. Ferrari, W. Kozlowska, A. Mairani and F. Ballarini (2019), First benchmarking of the BIANCA model for cell survival predictions in a clinical hadron therapy scenario, Physics in Medicine and Biology 64/21, 215008 19. Mario P Carante, Giulia Aricò, Alfredo Ferrari, Christian P Karger, Wioletta Kozlowska, Andrea Mairani, Paola Sala, and Francesca Ballarini. In vivo validation of the BIANCA biophysical model: Benchmarking against rat spinal cord RBE data. International Journal of Molecular Sciences, 21(11):3973, 2020.
19. Mario P. Carante, Alessia Embriaco, Giulia Aricò, Alfredo Ferrari, Andrea Mairani, Stewart Mein, Ricardo Ramos, Paola R. Sala and Francesca Ballarini, Biological effectiveness of He-3 and He-4 ion beams for cancer hadrontherapy: a study based on the BIANCA biophysical model. Physics in Medicine and Biology 66, 195009
20. Alessia Embriaco, Ricardo L. Ramos, Mario P. Carante, Alfredo Ferrari, Paola Sala, Valerio Vercesi and Francesca Ballarini (2021), Healthy tissue damage following cancer ion therapy: a radiobiological database predicting lymphocyte chromosome aberrations based on the BIANCA biophysical model.Int J Mol Sci 22(19), 10877
21. L Simonsen, T Slaba, P Guida, A Rusek, NASA's first ground-based Galactic Cosmic Ray simulator: enabling a new era in space radiobiology research. PLOS Biology 2020
22. F. Ballarini and A. Ottolenghi (2003), Chromosome aberrations as biomarkers of radiation exposure: modelling basic mechanisms. Adv. Space Res. 31(6), 1557
23. F. Ballarini, A. Ottolenghi (2004), A model of chromosome aberration induction and CML incidence at low doses. Radiat. Environ. Biophys. 43, 165-171.

A multi-detector experimental setup for the study of space radiation shielding materials: Measurement of secondary radiation behind thick shielding and assessment of its radiobiological effect

Felix Horst[1], *Daria* Boscolo[1], *Giorgio* Cartechini[2,3], *Marco* Durante[1,4], *Carola* Hartel[1], *Ekaterina* Kozlova[1], *Chiara* La Tessa[2,3], *Marta* Missiaggia[2,3], *Enrico* Pierobon[2,3], *Torsten* Radon[1], *Riccardo* Ridolfi[5,6], *Sylvia* Ritter[1], *Christoph* Schuy[1], *Alexey* Sokolov[1], *Uli* Weber[1], *Miroslav* Zbořil[7*]

[1]GSI Helmholtzzentrum für Schwerionenforschung Darmstadt, Planckstraße 1, 64291 Darmstadt, Germany
[2]Trento Institute for Fundamental Physics and Applications – Istituto Nazionale di Fisica Nucleare (TIFPA-INFN), Via Sommarive 14, 38122 Trento, Italy
[3]University of Trento, Via Sommarive 14, 38123 Trento, Italy
[4]Technical University of Darmstadt, Institute for Condensed Matter Physics, Hochschulstr. 6, 64289 Darmstadt, Germany
[5]Istituto Nazionale di Fisica Nucleare (INFN), Section of Bologna, Viale Berti Pichat 6/2, 40127 Bologna, Italy
[6]University of Bologna, Department of Physics and Astronomy, Viale Berti Pichat 6/2, 40127 Bologna, Italy
[7]Physikalisch-Technische Bundesanstalt (PTB), Bundesallee 100, 38116 Braunschweig, Germany

Abstract. Space agencies have recognized the risks of astronauts' exposure to space radiation and are developing complex model-based risk mitigation strategies. In the foundation of these models, there are still significant gaps of knowledge concerning nuclear fragmentation reactions which need to be addressed by ground-based experiments. There is a lack of data on neutron and light ion production by heavy ions, which are an important component of galactic cosmic radiation (GCR). A research collaboration has been set up to characterize the secondary radiation field produced by GCR-like radiation provided by a particle accelerator in thick shielding. The aim is to develop a novel method for producing high-quality experimental data on neutron and light ion production in shielding materials relevant for space radiation protection. Four complementary detector systems are used to determine the energy and angular distributions of high-energy secondary neutrons and light ions. In addition to the physical measurement approach, the biological effectiveness of the secondary radiation field is determined by measuring chromosome aberrations in human peripheral lymphocytes placed behind the shielding. The experiments are performed at the heavy ion

* Corresponding author: miroslav.zboril@ptb.de

synchrotron SIS18 at the GSI Helmholtzzentrum für Schwerionenforschung in Darmstadt, Germany, in the frame of FAIR Phase-0.

1 Motivation

The health effects of space radiation on astronauts represent a major limiting factor for long-duration human space missions beyond low Earth orbit [1]. The radiation exposure can potentially cause cancer and other late effects. National space agencies have recognized the risks related to exposure to space radiation and are developing complex model-based risk mitigation strategies [2, 3]. The space radiation risk models are based on interplay of radiation transport physics and radiobiology. In both fields, there are still significant gaps of knowledge which need to be closed by collecting high-quality experimental data and developing the corresponding theoretical models and simulation codes.

During passage through the shield, a major fraction of the heavy ions in the galactic cosmic radiation (GCR) break up due to nuclear fragmentation reactions into neutrons and light charged fragments. A relevant literature gap exists for neutron and light ion production in thick shielding upon the impact of GCR [4]. Due to their high penetration power and high biological effectiveness, neutrons represent a significant threat to the astronauts' health. The issue of neutron production in thick shielding is particularly important in design studies of habitats constructed from the local regolith, for instance for longer manned missions to the Moon or Mars surface. Recent comparisons of common particle transport codes used in the space radiation community (e.g., FLUKA, Geant4, MCNP, PHITS, HZETRN and SHIELD) suggest that differences in the neutron and light ion production models still cause significant discrepancies [2]. The reasons for that are uncertainties in the nuclear physics models describing the heavy ion total reaction cross sections [5] and the limited base of experimental double-differential cross sections [6].

To tackle these issues, a research collaboration has been set up to characterize the secondary radiation field produced by GCR-like radiation provided by a particle accelerator in thick shielding by means of a multi-detector setup consisting of complementary experimental systems. A cylindrical target is irradiated with a high-energy heavy ion beam and the various detectors, arranged around the target at defined angles, characterize different aspects of the secondary radiation field. The experiments at GSI Helmholtzzentrum für Schwerionenforschung in Darmstadt, Germany, are performed in the framework of the Investigations into Biological Effects of Radiation (IBER) program funded by the European Space Agency (ESA).

In the present article, the detection systems and their interplay are described. The different physical detector systems used are the Bonner sphere spectrometer NEMUS, an ambient dosimeter and a Bonner spectrometer based on thermoluminescence dosimetry (TLD), scintillator telescopes with time-of-flight (ToF) electronics and a tissue-equivalent proportional counter (TEPC). In addition, human blood lymphocytes are used as biological dosimeters. Based on the combination of these different methods, the secondary radiation field at the chosen angles of interest is well characterized. The quantities obtained with the different systems are neutron and light ion energy distributions, absorbed dose, ambient dose equivalent, microdosimetric spectra and the mean quality factor.

2 Experimental setups and methodology

A schematic of the experimental setup used during first experiments performed in Cave A at GSI with 1 GeV/u ^{56}Fe ions on a thick aluminum target is shown in Figure 1. As angles of interest 15° and 40° relative to the beam axis were chosen. As an example, the setup for irradiation of the human blood lymphocytes with simultaneous measurements using TEPC detectors on the other side of the beam axis is shown. The primary 1 GeV/u ^{56}Fe ion beam exits the vacuum beam line through a thin exit window and penetrates a beam monitor ionization chamber. The beam intensity is adjusted to match the optimal measurement conditions of each detector system. Optionally, for low intensity measurements like during the ToF experiment, also a plastic scintillator can be inserted. The monitor ionization chamber can be calibrated in terms of number of primary particles either by an absolute absorbed dose to water measurement in a scanned field under defined reference conditions [7] or by comparison of the ionization current with the rate measured by the plastic scintillator. After passing the beam monitors, the primary ions are stopped in a thick aluminum target (20 cm diameter) and produce different secondary particle species. The most abundant secondary particles are neutrons and protons. As indicated in Fig. 1, the distance of the different systems from the target center can be chosen individually to adapt to specific dose rate requirements. The difference in distances to the target center can, in principle, lead to different neutron spectra due to different contributions of the room-return neutrons. For the first evaluation of the physical detectors' and biodosimetry results, this effect will be assessed by a dedicated Monte Carlo study.

Aluminum is a "standard material" for space radiation protection research as it is the main structural material in space crafts and was therefore chosen for the first measurement campaign. For future experiments it is intended to characterize other materials as well, for example Moon regolith which could be used for lunar habitats [8], and to use also other primary ion species than ^{56}Fe ions.

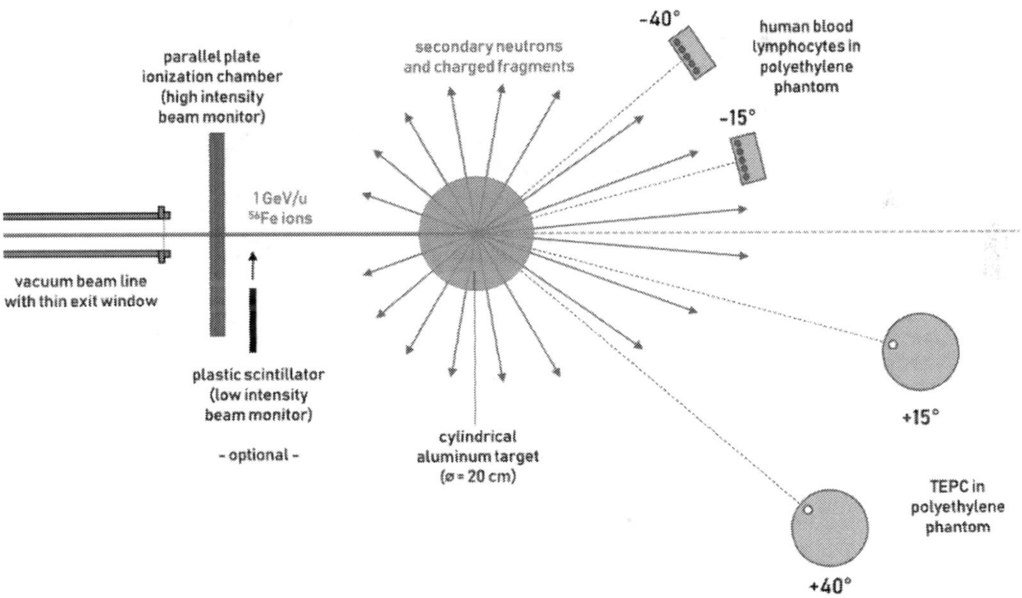

Fig. 1. Schematic of the multi-detector setup used for characterization of the secondary neutron and light ion radiation field produced by a high-energy heavy ion beam in a thick target.

In the following Sections, each experimental system used is described in more detail and the interplay of the different systems is depicted.

2.1 PTB Bonner sphere spectrometer NEMUS

A Bonner sphere spectrometer (BSS) is a commonly used detector system for the measurement of neutron energy distributions in unknown fields [9, 10]. A BSS consists of a set of moderating spheres with different diameters and a thermal neutron sensor that is placed at the center of each sphere. Each sphere plus thermal sensor combination has a different energy-dependent response to neutrons. The peak of the neutron response function shifts to higher neutron energies as the size of the moderator increases. It is common practice to measure also with the thermal neutron sensor without a moderating sphere (i.e., the bare detector).

In the framework of this project, the BSS NEMUS [11] developed at PTB in Braunschweig, Germany, is used. The BSS NEMUS is a secondary transfer standard of PTB, traceable to the primary standards of neutron fluence rate. It consists of ten polyethylene (PE) spheres with diameters ranging from 7.6 cm (3 inch) to 30.5 cm (12 inch). The set also contains a bare detector (diameter 3.2 cm), a Cd-covered detector, and four modified spheres with lead (Pb) and copper (Cu) shells. Thanks to these metal shells, embedded in the PE spheres, the response functions of the four modified spheres dramatically increase for neutron energies above 50 MeV. Typically, the central thermal neutron sensors (CTNS) are spherical ^{3}He-filled proportional counters (type SP9, company Centronic Ltd.), detecting the thermalized neutrons via the reaction ^{3}He(n,p)^{3}He with a Q value of 764 keV. For the measurements carried out at GSI, we use SP9 counters with ^{3}He pressure of 20 kPa and 200 kPa, respectively, to control the possible effects of dead time-caused signal losses. In addition, a ^{235}U-coated cylindrical fission chamber (type 307719, company LND, Inc.) is alternatively used as CTNS on certain measurement locations, to allow for a better discrimination of the neutron- and gamma-components of the secondary field behind the target. This is possible thanks to the neutron detection mechanism of the ^{235}U fission chamber (^{235}U(n,f)) with a large Q value of about 200 MeV. In addition, the thermal neutron detection efficiency of the specific ^{235}U fission chamber, used in this project, is by a factor of ~200 lower compared to the widely used ^{3}He-filled SP9 counter with 200 kPa filling pressure, which also improves the control on possible dead-time effects.

To cover the complete energy range of the expected neutron field and minimize the required measurement time, we chose a subset of NEMUS spheres, namely a bare counter, spheres of diameters of 7.6, 10.2, 12.7, 15.2, 20.3, 25.4 and 30.5 cm, respectively, plus all four modified spheres with metal shells. This way it is possible to cover the entire neutron energy spanning from the thermal neutrons up to the high-energy neutrons above 1 GeV. It should be noted, however, that the neutron response functions of the BSS NEMUS are validated in the PTB neutron reference fields only up to 20 MeV, and above this value the response functions rely purely on nuclear models implemented in the MCNP6 code. For quality-assurance reasons, the CTNS signals are recorded in a time-resolved manner. From the pulse-height spectra, the number of neutron-induced counts in each sphere is determined. This information is used in the unfolding process, together with the NEMUS response functions and an *a priori* (default) neutron energy distribution (spectrum) obtained from dedicated Monte Carlo simulations, to derive the resulting neutron spectrum.

2.2 TLD-based neutron ambient dosimeters and Bonner spectrometer

For neutron ambient dose measurements, the GSI ball neutron dosimeter [12] is used. This passive dosimeter, equipped with Harshaw TLD cards (one TLD-600H and one 700H pair), has a neutron response function close to fluence-to-ambient-dose conversion coefficients and, therefore, provides read-out values close to the ambient dose equivalent $H^{*}(10)$. First results for 1 GeV/u ^{56}Fe ions on a thick aluminum target have been published [13] and the dosimeter $H^*(10)$ readings were comparable with the predictions by the Monte Carlo code FLUKA.

Furthermore, a BSS-like spectrometer set based on the same TLD cards, consisting of one PE box ($2 \times 5 \times 7$ cm^3), one PE cylinder (diameter: 5.1 cm, height: 6 cm), four PE spheres (diameters of 12.7, 20.3, 25.4 and 30.5 cm, respectively), one PE sphere (diameter of 32.5 cm) with a 1 cm thick lead layer as well as one polyvinyl chloride (PVC) cylinder (height and diameter: 14 cm), also with additional lead and PE layers, is used for neutron spectrometry. For this purpose, the elements of the set are irradiated one after another at a certain position and the spectrum is unfolded from the readouts using the GRAVEL [14] unfolding program. First results for 1 GeV/u ^{56}Fe ions on aluminum have been published as well [15] and as for $H^*(10)$, also the measured neutron spectra are in reasonably good agreement with FLUKA calculations.

2.3 Time-of-flight spectrometry using scintillator telescopes

Another common method for measuring spectra of high-energy neutrons and charged fragments produced in heavy ion reactions is time-of-flight (ToF) spectrometry, where scintillators or combinations of scintillators ("scintillator telescopes") are used as particle detectors and to discriminate between different particle types [16]. The time between the heavy ions hitting the target and the particles arriving at the detector is recorded in event-by-event mode. From the time-of-flight of the single particles, velocity distributions are obtained which can then be converted into energy spectra of secondary neutrons and charged fragments (protons, deuterons, tritons, ^{3}He, ^{4}He, etc.).

While for charged particles, scintillators have practically a 100 % detection efficiency, the neutron efficiency that is reached for common scintillator materials and geometries is typically not higher than 15 %, depending, however, on the specific scintillator thickness. The neutron efficiency of a scintillator is usually a function of energy and this has to be taken into account to obtain absolute neutron spectra from such a measurement [17]. The lowest neutron energy that can be detected by the ToF method is around 10 MeV with common experimental electronics, therefore only the high-energy part of a neutron spectrum is determined.

2.4 Tissue-equivalent proportional counter

The distribution of energy deposition on a micrometer scale as relevant for mechanistic understanding of radiobiological effects can be measured with a tissue-equivalent proportional counter (TEPC). The microdosimetric quantities obtained in such a measurement are spectra of specific energy and lineal energy, the equivalents to the macroscopic quantities absorbed dose and linear energy transfer (LET).

The TEPC used in the experiments described in this article is the model LET-1/2 from the company Far West Technology. The TEPC detector is placed at 1 cm depth within a polyethylene phantom (see Fig. 1) to be consistent with the $H^*(10)$ measurements performed with TLDs and the irradiation of blood lymphocytes at the same depth. It is operated with a voltage of 700 V and read out by a charge-sensitive preamplifier (model A422A from CAEN). As the TEPC signals can span several orders of magnitude, they are fed into three different shaping amplifiers: two of model N968 from CAEN with gains of 100 and 1000, and one of model 7243E by Intertechnique with a gain of 10. Finally, the amplified signals are recorded by multichannel analyzers (models 926/927 from ORTEC). The three recorded spectra are overlapping so that they can be merged together in an offline analysis step.

The obtained microdosimetric spectra can then be analyzed to assess the radiation quality [18, 19].

2.5 Chromosome aberrations in human blood lymphocytes (biodosimetry)

All detector systems described above are physical detectors. A complementary approach to assess the biological effect of a radiation exposure is the so-called cytogenetic biodosimetry, i.e., the analysis of chromosome aberrations in human peripheral blood lymphocytes. Chromosome aberrations are a sensitive indicator of an exposure to ionizing radiation and provide information on radiation-associated health risks, in particular the risk of cancer induction [20]. In the scope of space radiation protection, a comparable approach using human fibroblasts was presented by Slaba *et al.* [21].

Within this project, biodosimetry is performed with human peripheral blood lymphocytes following standard techniques [22]. Blood is taken from a healthy donor. Then, the lymphocytes are isolated, transferred into plastic tubes and placed at a depth of 1 cm within a polyethylene phantom. Samples are irradiated with doses from about 100 mGy up to 1 Gy. The absorbed dose is measured using a PTW 30013 Farmer ionization chamber that is placed in the phantom next to the cell samples, enabling a direct comparison of biological effects and physical dose. After exposure, the cytogenetic damage is analyzed in lymphocytes at the first post-irradiation mitosis. As the difference in the energy deposition of high- and low-LET radiation does not only affect the number of aberrations produced per unit dose, but also the types of aberrations, the high-resolution 24-color karyotyping technique (multiplex fluorescence *in situ* hybridization, mFISH) is applied. This method allows, for example, to visualize complex chromosome exchanges that are a signature of high-LET radiation [23, 24]. As previous measurements were mostly done with fission neutrons [25] and there are hardly any high-resolution data for high-energy neutrons as produced by GCR, the presented experiments are expected to fill this gap.

The irradiations are performed with high beam intensity to reach doses above 100 mGy in the scattered radiation field behind the target within a reasonable time (i.e., a few hours). Taking into account that the dose rate under these irradiation conditions is rather low (a few mGy/min), dose-effect curves for the induction of aberrations in lymphocytes by 220 kV X-rays (i.e., the reference radiation) at a comparable dose rate will be measured as well.

3 Interplay between experimental setups

The presented multi-detector approach allows a very accurate characterization of the secondary radiation fields studied. The single systems suffer from different issues; however, these can be overcome by combining the different measurements that complement each other in that sense. For instance, some of the Bonner spectrometer elements, both in the NEMUS setup and in the TLD-based system, respond not only to neutrons but also to high-energy protons and other charged fragments. Therefore, information about the secondary charged fragment spectra is required to correct the detector readings, and this is available from the scintillator ToF measurements.

With the ToF spectrometry, it is possible to determine the neutron spectra only above 10 MeV, while the energy distribution of the secondary neutrons is continuous and extends down to thermal neutron energies. With the Bonner spectrometry methods, the entire neutron spectra are determined, but with lower resolution at high energies. Therefore, overlapping the neutron spectra obtained by ToF and Bonner spectrometry can give a more accurate measurement of the neutron spectra than both methods alone. In addition, the absolute scale of the neutron energy distributions, resulting from the analysis of ToF, BSS NEMUS and TLD-based BSS, will be systematically compared to achieve a unified assessment of the absolute secondary neutron production yields.

Neutrons show an increased radiobiological effectiveness, i.e., a larger biological effect (e.g., induce more chromosome aberrations [25]) compared to low-LET reference radiation at the same absorbed dose level. This higher relative biological effectiveness (RBE) is due to the secondary particles (mostly recoil protons) generated by neutrons when interacting with tissue. The magnitude of these effects depends strongly on the neutron energy spectrum as well as on the tissue type, dose and dose rate. In radiation protection, the maximum RBE for oncogenic transformations of a given radiation field is approximated by so-called quality factors Q. The International Commission on Radiological Protection (ICRP) has defined a function which describes the dependence of the quality factor Q on the LET [26] which can also be applied to neutrons if the LET spectrum of their secondary particles is known [27]. The absorbed dose D multiplied with Q gives the dose equivalent H which can serve as a measure for the late risk associated with a radiation exposure. The lineal energy spectra obtained with the TEPC detector are a good proxy for the LET spectra and can be used to calculate the mean quality factor using the ICRP function. The absorbed dose D is directly measured with Farmer ionization chambers during irradiation of the biological samples and the neutron dose equivalent H is measured by the TLD-based ambient dosimeters. From these two quantities, the mean quality factor can be calculated as well. The RBE values, obtained from the biodosimetry, will be compared to the quality factors obtained by physical dosimetry, to understand how reliably the quality factors, used in radiation protection practice, can actually estimate the biological effectiveness of mixed radiation fields with a high neutron component induced by GCR-like heavy ions in thick shielding, as relevant for space radiation protection.

4 Summary and Outlook

A multi-detector setup for characterization of secondary neutron and light ion fields, produced by GCR-like heavy ions, which is of interest for the field of space radiation protection, was presented. The different experimental methods were described, and their interplay was outlined.

First experiments with 1 GeV/u ^{56}Fe ions on a thick aluminum target have been carried out in Cave A at GSI Helmholtzzentrum für Schwerionenforschung in Darmstadt, Germany, in the framework of the IBER program funded by ESA. The results obtained with the TLD-based detector systems are already published [13, 15] and the measurement data from the other methods are currently being analyzed.

Future experiments will also study different materials like Moon regolith and other primary ion species.

This experiment is performed at the heavy ion synchrotron SIS18 at the GSI Helmholtzzentrum für Schwerionenforschung, Darmstadt (Germany) in the frame of FAIR Phase-0. The beam time is funded by the European Space Agency (ESA) in the framework of the IBER project. This experiment is funded by the German Space Agency at the German Aerospace Center (DLR) under the funding number 50WB2125. The authors would like to thank M. Dommert, T. Klages, A. Lücke, M. Reginatto and A. Zimbal from PTB Braunschweig for fruitful cooperation.

References

1. M. Durante, Life Sciences in Space Research **1**, 2 (2014)
2. J.W. Norbury, T.C. Slaba, S. Aghara, F.F. Badavi, S.R. Blattnig, M.S. Clowdsley, L.H. Heilbronn, K. Lee, K.M. Maung, C.J. Mertens, J. Miller, R.B. Norman, C.A. Sandridge, R. Singleterry, N. Sobolevsky, J.L. Spangler, L.W. Townsend, C.M. Werneth, K. Whitman, J.W. Wilson, S.X. Xu, C. Zeitlin, Life Sciences in Space Research **22**, 98 (2019)

3. L. Walsh, U. Schneider, A. Fogtman, C. Kausch, S. McKenna-Lawlor, L. Narici, J. Ngo-Anh, G. Reitz, L. Sabatier, G. Santin, L. Sihver, U. Straube, U. Weber, M. Durante, Life Sciences in Space Research **21**, 73 (2019)
4. J.W. Norbury, T.C. Slaba, Life Sciences in Space Research **3**, 90 (2014)
5. F. Luoni, F. Horst, C.A. Reidel, A. Quarz, L. Bagnale, L. Sihver, U. Weber, R.B. Norman, W. de Wet, M. Giraudo, G. Santin, J.W. Norbury, M. Durante, New J. Phys. **23**, 101201 (2021)
6. J.W. Norbury, G. Battistoni, J. Besuglow, L. Bocchini, D. Boscolo, A. Botvina, M. Clowdsley, W. de Wet, M. Durante, M. Giraudo, T. Haberer, L. Heilbronn, F. Horst, M. Krämer, C. La Tessa, F. Luoni, A. Mairani, S. Muraro, R.B. Norman, V. Patera, G. Santin, C. Schuy, L. Sihver, T.C. Slaba, N. Sobolevsky, A. Topi, U. Weber, C.M. Werneth, C. Zeitlin, Frontiers in Physics **8**, 565954 (2020)
7. F. Luoni, U. Weber, D. Boscolo, M. Durante, C.-A. Reidel, C. Schuy, K. Zink, F. Horst, Frontiers in Physics **8**, 568145 (2020)
8. J. Miller, L. Taylor, C. Zeitlin, L. Heilbronn, S. Guetersloh, M. DiGiuseppe, Y. Iwata, T. Murakami, Radiat. Meas. **44**, 163 (2009)
9. D.J. Thomas, A.V. Alevra, Nucl. Instrum. Meth. A **476**, 12 (2002)
10. A.V. Alevra, D.J. Thomas, Radiat. Prot. Dosim. **107**, 37 (2003)
11. B. Wiegel, A.V. Alevra, Nucl. Instrum. Meth. A **476**, 36 (2002)
12. G. Fehrenbacher, F. Gutermuth, E. Kozlova, T. Radon, R. Schuetz, Radiat. Prot. Dosim. **125**, 209 (2007)
13. D. Boscolo, D. Scognamiglio, F. Horst, U. Weber, C. Schuy, M. Durante, C. La Tessa, E. Kozlova, A. Sokolov, I. Dinescu, T. Radon, D. Radeck, M. Zbořil, Frontiers in Physics **8**, 365 (2020)
14. M. Matzke, Radiat. Prot. Dosim. **107**, 155 (2003)
15. A. Sokolov, E. Kozlova, D. Boscolo, M. Durante, F. Horst, T. Radon, C. Schuy, U. Weber, J. Instrum. **16**, P10022 (2021)
16. T. Nakamura, L. Heilbronn, *Handbook on Secondary Particle Production and Transport by High-Energy Heavy Ions*. World Scientific Publishing Co. Pte. Ltd., ISBN 981-256-558-2 (2006)
17. K. Gunzert-Marx, D. Schardt, R.S. Simon, F. Gutermuth, T. Radon, V. Dangendorf, R. Nolte, Nucl. Instrum. Meth. A **536**, 146 (2005)
18. G. Martino, M. Durante, D. Schardt, Phys. Med. Biol. **55**, 3441 (2010)
19. M. Missiaggia, G. Cartechini, E. Scifoni, M. Rovituso, F. Tommasino, E. Verroi, M. Durante, C. La Tessa, Phys. Med. Biol. **65**, 245024 (2020)
20. S. Bonassi, H. Norppa, M. Ceppi, U. Strömberg, R. Vermeulen, A. Znaor, A. Cebulska-Wasilewska, E. Fabianova, A. Fucic, S. Gundy, I.-L. Hansteen, L.E. Knudsen, J. Lazutka, P. Rossner, R.J. Sram, P. Boffetta, Carcinogenesis **29**, 1178 (2008)
21. T.C. Slaba, I. Plante, A. Ponomarev, Z.S. Patel, M. Hada, Radiat. Res. **194**, 246 (2020)
22. International Atomic Energy Agency: *Cytogenetic Dosimetry: Applications in Preparedness for and Response to Radiation Emergencies*, IAEA Report, Vienna, Austria (2011)
23. R. Lee, S. Sommer, C. Hartel, E. Nasonova, M. Durante, S. Ritter, Mutat. Res. - Gen. Tox. En. **701**, 52 (2010)
24. R.M. Anderson, Clin. Oncol. **31**, 311 (2019)

25. L.C. Paterson, A. Yonkeu, F. Ali, N.D. Priest, D.R. Boreham, C.B. Seymour, F. Norton, R.B. Richardson, Radiat. Res. **195**, 211 (2020)
26. International Commission on Radiological Protection (ICRP): *Recommendations of the International Commission on Radiological Protection,* ICRP Publication 60 (1990)
27. H. Schuhmacher, B.R.L. Siebert, Radiat. Prot. Dosim. **40**, 85 (1992)

Elemental analysis of concrete via fast neutron transmission and scattering spectrometry

Tanya Hutton[1*], *Andy* Buffler[1], and *Mark* Alexander[2]

[1]Metrological and Applied Sciences University Research Unit (MeASURe), Department of Physics, University of Cape Town, South Africa.
[2]Concrete Materials and Structural Integrity Research Unit (CoMSIRU), Department of Civil Engineering, University of Cape Town, South Africa.

Abstract. We report on the development of neutron-based techniques to non-destructively measure the composition of concrete. Previous experimental studies demonstrated the viability of the unfolding technique to determine the ratio of water, sand and cement in well-characterised concrete samples from the transmitted neutron energy spectrum. In this work, we used MCNP6 simulations to demonstrate the extension of the technique to determine elemental compositions from transmitted, or scattered neutron energy spectra. In both cases, the simulated energy spectra provided a reliable method to unfold the composition of samples with known elemental ratios. The precision of the technique was limited by the statistical uncertainties of the simulated spectra, particularly for the case of scattered neutrons. The accuracy of the technique was heavily dependent on the uniqueness of each of the elemental responses, and reasonable prior knowledge of the composition. Given the promising results at this stage, future developments will include the addition of further elements to the response matrix, and experimental verification.

1 Introduction

Hydrogen-rich materials offer excellent neutron shielding, and concrete is widely used in nuclear facilities due to its structural strength, low cost, low activation and low maintenance requirements. The types of concrete used in nuclear installations are subject to a series of codes and standards, such as ACI 318, ACI 349 and ACI 359, which apply to concrete in nuclear power plants though these are subject to ongoing research and development [1]. The nuclear regulator for the region is then responsible for ensuring compliance between the regulatory radiation safety standards and the physical installation at all stages of the life-cycle. In this project we aim to develop and characterize a new experimental research and testing facility that will be used to facilitate investigations requiring radiation-based characterization of concrete (and other materials) for the nuclear industry. Of particular interest is the ability to non-destructively determine the water content in concrete, which is a known problem for ageing nuclear facilities as a reduction in hydrogen content results in a

* Corresponding author: tanya.hutton@uct.ac.za

reduction in the effectiveness as a neutron shield [2]. In the case of radiation transport simulations of nuclear facilities, poor knowledge of the concrete composition, in particular the hydrogen content, can cause significant discrepancies between measured and simulated neutron spectra and associated dose rates [3].

Materials analysis utilising fast neutrons is advantageous in this context as fast neutrons are non-invasive, highly penetrating, sensitive to low mass nuclei and produce characteristic secondary radiation for each nuclide. In fast neutron transmission measurements, a beam of fast neutrons is incident upon a sample and the fluence, and potentially energy distribution, of the exiting neutrons are measured. Information concerning the neutrons which have interacted in the sample can be obtained if the energy dependent fluence entering the sample is known. This technique has been applied in many contexts, such as the detection of explosives [4] using a ns-pulsed neutron beam, or scanning for illicit materials in cargo by considering the ratio of neutron and gamma ray attenuation coefficients [5].

In previous studies at the University of Cape Town, the water content in concrete has been experimentally measured through spectral unfolding [6]. Energy dependent response functions were measured for water, sand and cement, and were successfully used to unfold the composition of a concrete sample within 5% of the known values. However, there may be instances where the composite ingredients are not available, and the material composition is often location specific. For example, sand mined from coastal regions will contain variable proportions of calcium carbonate relative to sand obtained from an inland mine [7]. In this work we investigate the use of fast neutron transmission and scattering spectrometry to characterise the elemental composition of materials through Monte Carlo radiation transport simulations.

2 Approach

There are three main signatures used for the analysis of materials in bulk using fast neutrons: transmitted neutrons; elastically and inelastically scattered neutrons; and gamma rays produced from neutron-induced reactions. As neutrons interact with atomic nuclei in distinctive, energy dependent ways, the signatures produced by each element are also distinct. If then, measurements of these signatures are made for a composite material, the elemental composition can be determined from these signatures according to Equation 1, where S_j is the measured signature for the composite sample in channel j, R_{jk} is the response function associated with element, k in channel j, and p_k is the projected number density of element, k.

$$S_j = \sum_k R_{jk} p_k \tag{1}$$

The elemental composition can then be de-convolved from the measured signatures through a least-squares minimisation procedure. To develop elemental response functions and verify the procedure, Monte Carlo radiation transport simulations of a highly simplified scenario were undertaken with MCNP6.1 [8] for the exemplar case of sand, a key constituent of concrete.

2.1 Simulation details

The simulations comprised of a 0.8 cm diameter pencil beam of fast neutrons, with a uniform energy distribution between 1.0 MeV and 14.0 MeV, incident upon elementally, and isotopically, pure samples of ^{1}H, ^{16}O, ^{28}Si, and ^{40}Ca. All neutron reaction cross sections used in these simulations were from the ENDF/B-VII.1 evaluation [9]. The cylindrical sample dimensions were defined as 1.0 cm along the beam axis, with a radius of 1.0 cm, with a fixed atom density of 1.0 b^{-1}cm^{-1}. Four spherical detector regions with radii of 1.0 cm were located at 0°, 45°, 90° and 135° relative to the beam axis, at a distance of 50.0 cm from the centre of the sample. These detector regions were used to determine the average energy dependent neutron flux using a track-length estimator (F4 tally). In this instance no additional variance reduction was implemented, and each simulation was run such that the statistical uncertainties on each bin were below 10 %. These simulations are highly idealised in terms of the incident neutron spectrum and sample properties for demonstration purposes. To realise this process experimentally, a broad spectrum neutron source such as ^{241}Am^{9}Be, and variable sample properties, could be used with appropriate normalisation.

2.2 Construction of response functions

Simulations of the elemental samples were used to develop the response functions required for the unfolding process. The scattered neutron spectra (45° - 135°) provide information on the neutrons which have interacted within the sample, so these can be directly used as the elemental response R_{jk} in this instance as described in Equation 1.

However, in the case of the transmitted neutron energy spectrum, this only provides information on the neutrons which have not interacted within the sample and requires further manipulation to be utilised as a response. Here we use the effective removal cross section [10], Σ^{R}_{jk}, calculated according to Equation 2, where ϕ_{jk} is the transmitted neutron flux in energy bin j for element k and sample thickness t, and ϕ_j^0 is the incident neutron flux in energy bin j. For composite materials, the effective removal cross section Σ_j^R is a linear combination of it constituent elements according to Equation 3, where p_k is the projected number density of the constituent element k, with density ρ_k.

$$\Sigma^{\mathrm{R}}_{jk} = -\ln\left(\frac{\phi_{jk}}{\phi_j^0}\right)\frac{1}{t} \tag{2}$$

$$\Sigma^{\mathrm{R}}_{\mathrm{j}}/\rho = \sum_k p_k(\Sigma^R_{jk}/\rho_k) \tag{3}$$

2.3 Unfolding of elemental composition of known materials

In order to determine the relative amounts of each element within a composite material, an unfolding procedure was utilised, in this instance the multi-channel unfolding programs GRAVEL and MAXED which form part of the UMG-33 software package [11].

To verify the unfolding procedure, simple combinations of hydrogen, oxygen, silicon and calcium were simulated, as given in Table 1, with the same geometry as used previously. The sand analogue used in these simulations is loosely based on the composition obtained from XRF analyses of real samples [6]. The simulated response functions for the elemental samples were used to deconvolve the ratios of the different elements in the composite media from the transmitted and scattered signatures. A two-step unfolding process was implemented, first using GRAVEL, which is based on the SAND-II algorithm, and produces a solution spectrum through least-squares minimisation. As the starting point for the unfolding process, equal ratios of all elements were assumed, effectively representing the worst case scenario with regards to prior knowledge of the composition. The solution ratios produced by GRAVEL were then used as *a priori* information for unfolding with MAXED, which uses a maximum entropy approach, and allows for the propagation of uncertainties.

Table 1. Composite sample details.

Sample	**ρ (g cm^{-3})**
H_2O	1.0
SiO_2	2.7
CaO	3.3
Sand 1:8:1 (H_2O:SiO_2:CaO)	1.5

3 Results & Analysis

Each elemental simulation was run with a variable number of particle histories in order to achieve convergence, and comparable statistics for each of the simulated signatures. Figure 1 shows the spatial distribution of neutrons for (a) hydrogen and (b) silicon targets. As expected, the scattered neutron field is more isotropic for hydrogen relative to the forward biased distribution for silicon.

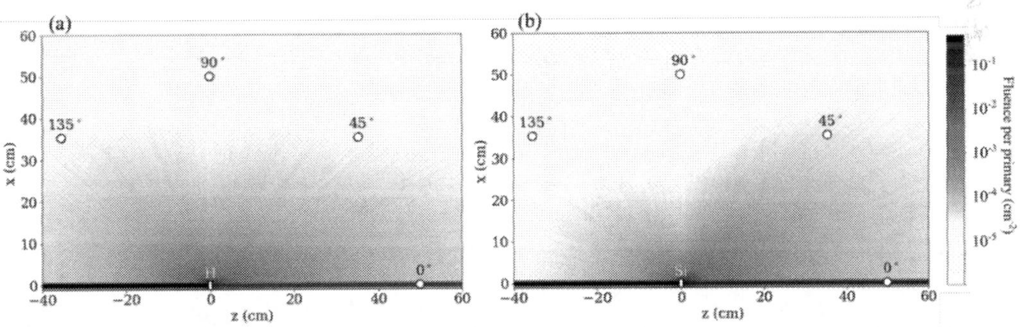

Fig. 1. Total neutron fluence over the simulated geometry for a broad spectrum neutron beam along the z-axis, incident on targets comprised of (a) hydrogen and (b) silicon respectively. Detector regions are indicated with the white circles.

3.1 Construction of response functions

Within each of the detector regions, the energy dependent neutron fluence was scored, and can be seen in Figure 2. For clarity, statistical uncertainties have been excluded from the figures. For the transmitted (0°) spectra, statistical uncertainties were below 1 %, and between

1 % and 10 % for the scattered spectra (45°-135°). Unique features in the simulated spectra, such as the smoothly varying distribution for hydrogen, and the enhancement in the 2-3 MeV region for oxygen, start to become apparent and are directly related to the underlying interaction cross sections.

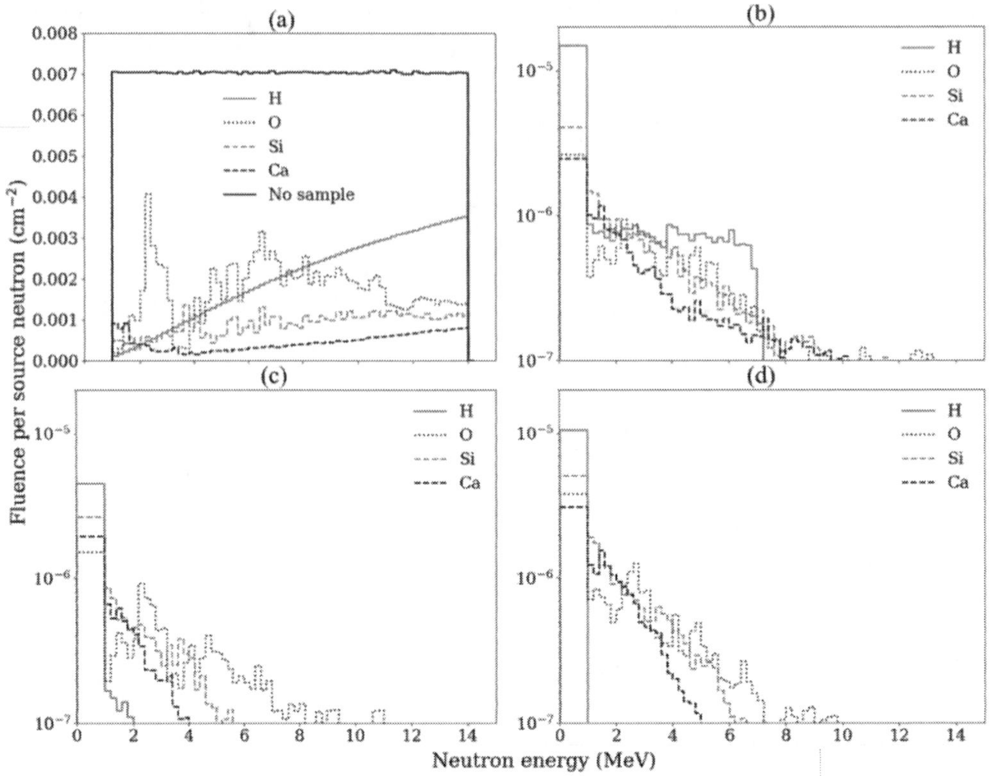

Fig. 2. Energy dependent neutron fluence, per source neutron, for elemental samples in the detector regions at (a) 0°, (b) 45°, (c) 90° and (d) 135° relative to the beam axis. Note that a log fluence scale is used for the scattered spectra (b-d).

From the simulated energy spectra the elemental response functions are constructed, utilising the effective removal cross section for the transmission simulations according to Equation 2, and selected energy ranges for the scattered spectra. The statistical uncertainties associated with the response functions were accounted for through an appropriate choice of precision. The elemental response functions were compiled into a response matrix with a format compatible with the UMG unfolding package, and can be seen in Figure 3.

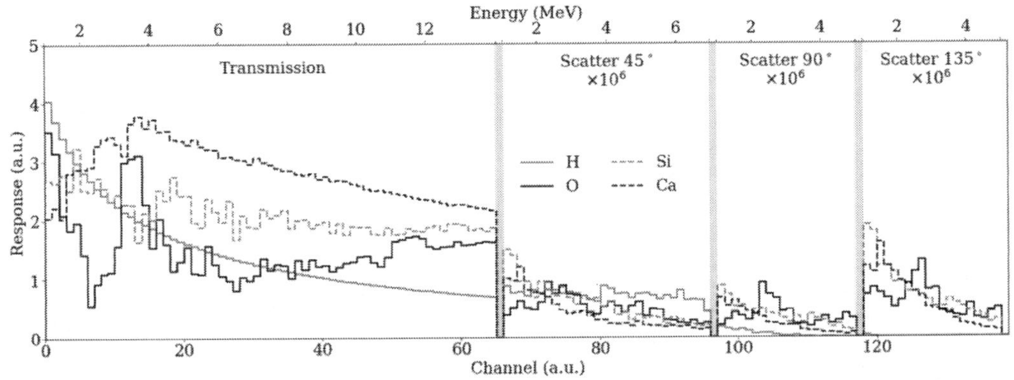

Fig. 3. Response functions constructed for elemental samples of hydrogen, oxygen, silicon and calcium. Effective removal cross sections are used between 1.0 MeV and 14.0 MeV for the transmission component. The scattered components consist of the simulated energy spectra between 1.0 MeV and 7.0 MeV for 45°, and 1.0 MeV and 5.0 MeV for 90° and 135°. For convenience the scattered fluence spectra have been multiplied by a factor of 10^6.

3.2 Unfolding of elemental composition of known materials

Transmitted and scattered energy spectra were obtained for the composite samples detailed in Table 1, and the simulated signatures shown in Figure 4 were constructed by the same process used to develop the elemental responses. The ratios of H, O, Si and Ca were unfolded from either the transmitted, or scattered signatures.

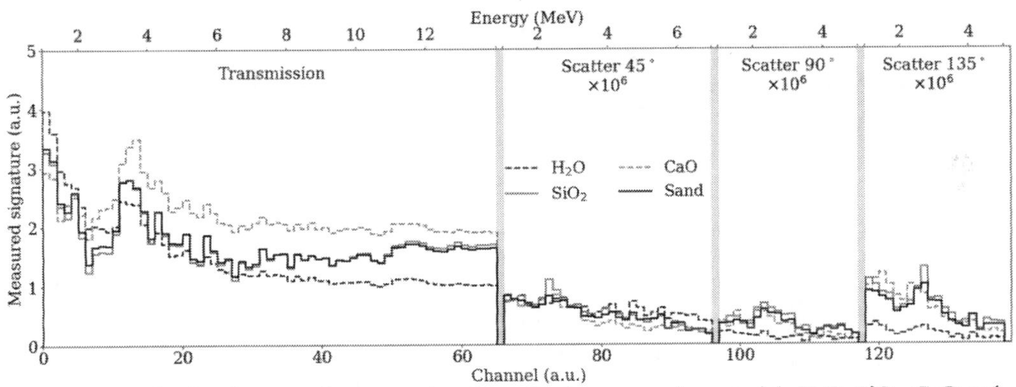

Fig. 4. Transmitted and scattered neutron signatures for the composite materials H_2O, SiO_2, CaO and sand as defined in Table 1. Statistical uncertainties have been excluded for clarity, but are below 1 % for the transmission signatures, and below 10 % for the scattered signatures.

For the exemplar case of H_2O, the simulated and re-folded signatures are shown in Figure 5. For the transmitted neutrons there is excellent agreement between the two over the full energy range with little deviation. In the case of scattered neutrons the quality of the calculated signature is limited by the poor statistics associated with the simulated signatures and a non-optimised energy binning structure, though the overall agreement is considered acceptable across the three angles used.

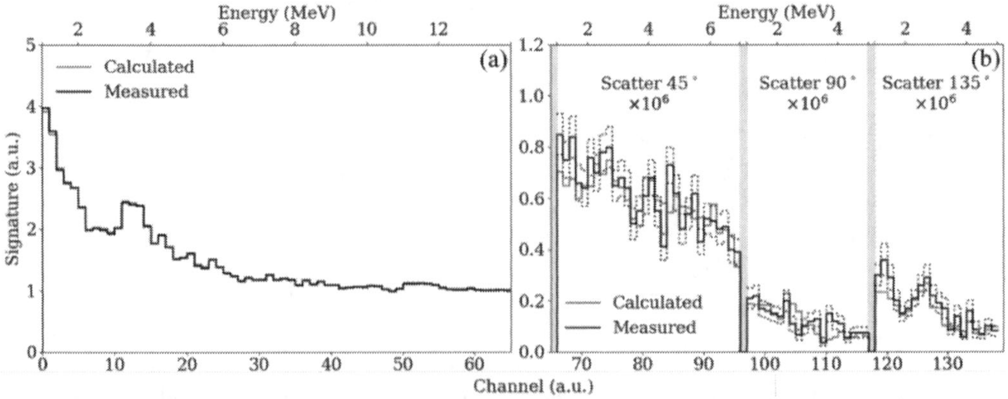

Fig. 5. Simulated signatures for (a) transmitted (0°) neutrons, and (b) scattered neutrons (45°-135°), shown together with the refolded signatures obtained using the MAXED package. Statistical uncertainties on the simulated signatures are indicated with the dotted lines.

The resulting elemental ratios obtained from the unfolding process of the composite materials are shown in Figure 6. In all but a few cases, the resulting elemental composition was in agreement with the known values. The most notable instance is the minor misclassification of calcium in CaO as silicon in the transmitted analysis, which suggests there is insufficient uniqueness between the two signatures with the current structure. This could be mitigated through a more appropriate choice of the initial ratios, but requires some prior knowledge of the material. For H_2O and sand, the ratio of hydrogen was reproduced from the unfolding analyses within the tolerance of the statistical uncertainties. The difference between the expected and unfolded elemental ratios for sand are presented in Table 2 with their statistical uncertainties.

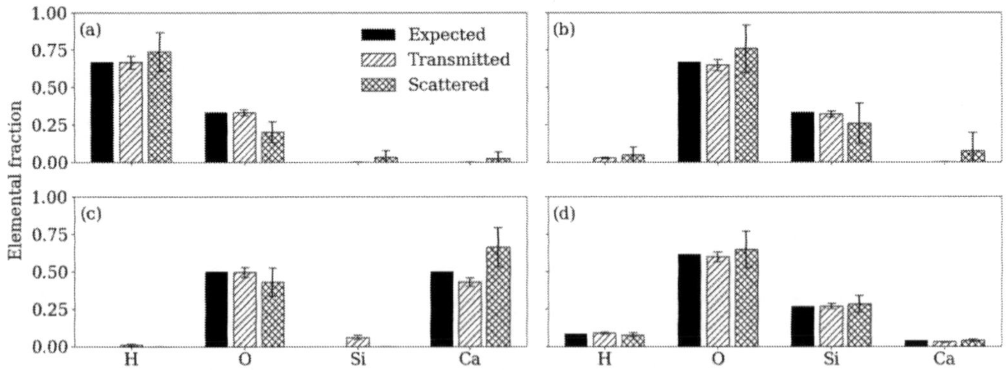

Fig. 6. Elemental fractions of H, O, Si and Ca unfolded from either the transmitted or scattered signatures for (a) H_2O, (b) SiO_2, (c) CaO, and (d) sand. In each case, the known ratio (black) is included for comparison.

Table 2. Difference between expected and unfolded elemental ratios for transmitted (Δ_T) and scattered (Δ_S) analyses for the case of sand.

	Δ_T (%)	Δ_S (%)
H	0.50 ± 0.55	-1.0 ± 1.0
O	-0.4 ± 3.1	1.2 ± 8.7
Si	0.5 ± 1.4	0.4 ± 4.0
Ca	-0.83 ± 0.27	0.05 ± 0.50

4 Conclusion

In this work we aimed to demonstrate the use of spectrum unfolding as a means to determine the elemental composition of materials. Monte Carlo simulations were used to produce energy dependent elemental response functions for transmitted and scattered neutrons. Elemental ratios were unfolded from simulated signatures for a series of composite materials, where the truth values were known. Practically, transmitted neutron spectroscopy is the preferred mode due to the higher detection rates, though this requires knowledge of the incident spectrum. Scattered neutron spectroscopy, while limited by potentially poor statistics, provides a viable alternative in instances where transmission is not possible, for example, if no safe access is possible behind a thick radiation shield. In all instances, the reliability of the unfolded elemental ratios is dependent on the quality and uniqueness of the response functions. Future developments on this project consist of the experimental verification of the response functions as far as reasonably possible, and the inclusion of additional elements to broaden the range of potential applications.

We thank the National Nuclear Regular (South Africa) through the Centre for Nuclear Safety and Security for their support of this project (CNSS0117-E5-UCT)

References

1. C. Ferraris, *Concrete Codes and Standards for Nuclear Power Plants: Recommendations for Future Development*, ANSI publications (2011)
2. E.G. Peterson, *Shielding Properties of Ordinary Concrete as a Function of Temperature*, USAEC, Report HW-65572 (1960)
3. M. Petit, Nucl. Sci. Eng., **195**:8, 846 (2021)
4. J.C. Overley, Appl. Radiat. Isot. **36:**3 185 (1985)
5. J.E. Eberhardt, S. Rainey, R.J. Stevens, B.D. Sowerby, J.R.Tickner, Appl. Radiat. Isot., **63**:2, 179 (2005)
6. A. Buffler, T. Hutton, T. Leadbeater, Int. J. Mod. Phys., **50**, 2060015 (2020)
7. B. Walker, *Fine Aggregate Resources in the Greater Cape Town Area*, Master's thesis, University of Cape Town (2013)
8. T. Goorley, M. James, T. Booth, F. Brown, J. Bull, L. J. Cox, J. Durkee, J. Elson, M. Fensin, R. A. Forster, J. Hendricks, H. G. Hughes, R. Johns, B. Kiedrowski, R. Martz, S. Mashnik, G. McKinney, D. Pelowitz, R. Prael, J. Sweezy, L. Waters, T. Wilcox, T. Zukaitis, Nucl. Tech., **180**, 298 (2012)
9. M.B. Chadwick, M. Herman, P. Obložinský, M.E. Dunn, Y. Danon, A.C. Kahler, D.L. Smith, B. Pritychenko, G. Arbanas, R. Arcilla, R. Brewer, D.A. Brown, R. Capote, A.D. Carlson, Y.S. Cho, H. Derrien, K. Guber, G.M. Hale, S. Hoblit, S. Holloway, T.D. Johnson, T. Kawano, B.C. Kiedrowski, H. Kim, S. Kunieda, N.M. Larson, L. Leal, J.P. Lestone, R.C. Little, E.A. McCutchan, R.E. MacFarlane, M. MacInnes, C.M. Mattoon, R.D. McKnight, S.F. Mughabghab, G.P.A. Nobre, G. Palmiotti, A. Palumbo, M.T. Pigni, V.G. Pronyaev, R.O. Sayer, A.A. Sonzogni, N.C. Summers, P. Talou, I.J. Thompson, A. Trkov, R.L. Vogt, S.C. van der Marck, A. Wallner, M.C. White, D. Wiarda, P.G. Young, Nucl. Dat. Sh., **112**:12, 2887 (2011)
10. R.D. Albert, T.A. Welton, *A simplified theory of neutron attenuation and its application to reactor shield design*, USAEC, Report WAPD-15 (1950)
11. M. Reginatto, Radiat. Meas., **45**, 1323 (2010)

Radiation exposure of microorganisms living in radioactive mineral springs

Sofia Kolovi[1a*], *Giovanna-Rosa* Fois[1], *Sarra* Lanouar[1], *Patrick* Chardon[1a], *Didier* Miallier[1], *Guillaume* Rivrais[1a], *Elisabeth* Allain[2a], *Lory-Anne* Baker[2a-3a], *Celine* Bailly[4a], *Aude* Beauger[2a], *David-Georges* Biron[3a], *Yihua* He[4a], *Guillaume* Holub[5a], *Anne-Helene* Le Jeune[3a], *Clarisse* Mallet[3a], *Herve* Michel[6a], *Gilles* Montavon[4a], *Benoit* Schoefs[7a], *Claire* Sergeant[5a], *Lydia* Maigne[1a], and *Vincent* Breton[1a]

[1] Laboratoire de Physique de Clermont-Ferrand (LPC) / CNRS IN2P3, Université Clermont Auvergne, 63178, Aubière, France
[2] Laboratoire de Géographie Physique et Environnementale (GEOLAB) / CNRS, Université Clermont Auvergne, 63057, Clermont-Ferrand, France
[3] Laboratoire Microorganismes : Génome et Environnement (LMGE) / CNRS, Université Clermont Auvergne, 63178, Aubière, France
[4] Laboratoire de physique subatomique et des technologies associées (SUBATECH) / CNRS IN2P3, Université de Nantes, 44307, Nantes, France
[5] Centre d'Etudes Nucléaires de Bordeaux-Gradignan (CENBG) / CNRS IN2P3, Université Bordeaux, 33170, Gradignan, France
[6] Institut de Chimie de Nice (ICN) / CNRS, 06108, Nice, France
[7] Métabolism, molecular engineering of Microalgae and applications (MIMMA), Biologie des Organismes Stress Santé Environnement, Le Mans University,72000, Le Mans, France
[a] Members of Long-Term Socio-Ecological Research (LTSER) «Zone Atelier Territoires Uranifères (ZATU)»

Abstract. The TIRAMISU collaboration gathers expertise from biologists, physicists, radiochemists and geologists within the Zone-Atelier Territoires Uranifères (ZATU) in France to analyze the radiation exposure of microorganisms living in naturally radioactive mineral springs. These springs are small waterbodies that are extremely stable over geological time scales and display different physicochemical and radiological parameters compared to their surroundings. Water and sediment samples collected in 27 mineral springs of the volcanic Auvergne region (Massif Central, France) have been studied for their microbial biodiversity and their radionuclide content. Among the microorganisms present, microalgae (diatoms), widely used as environmental indicators of water quality, have shown to display an exceptional abundance of teratogenic forms in the most radioactive springs studied (radon activity up to 3700 Bq/L). The current work presents a first assessment of the dose received by the diatoms inhabiting these ecosystems. According to ERICA tool, microorganisms living in most of the sampled mineral springs were exposed to dose rates above 10 μGy/h due to the large concentration of radium in the sediments (up to 50 Bq/g). Radiological analyses of water and sediments were used as inputs to Monte Carlo simulations at micro- (GATE) and nano- (Geant4-DNA) scale in order to assess the direct and indirect damages on the diatom DNA.

* Corresponding author: sofia.kolovi@clermont.in2p3.fr

1 Introduction

Naturally radioactive mineral springs can be described as isolated ecosystems of various physicochemical properties with elevated levels of radionuclide content in comparison to their surroundings in which minerals and radioisotopes of the three natural decays series (^{238}U, ^{232}Th and ^{235}U) are abundant [1,2]. Specifically, ^{226}Ra ($t_{1/2}$ = 1600 y) and its gaseous descendant ^{222}Rn ($t_{1/2}$ = 3.82 d), both emitting alpha particles of maximum energy 4.8 MeV and 5.5 MeV respectively [3], are found in high concentrations in the sediment and the water of the mineral springs. Recent studies concerning organisms that live in naturally radioactive environments have already shown a correlation between mutation rates and radioactivity [4].

Auvergne, situated in Massif Central of France, is a volcanic region rich in mineral springs and high uranium content [5]. Diatoms, a species of unicellular microalgae, thrive in these peculiar environments [6]. It has been recently observed that they display an exceptional abundance of teratogenic forms in the most radioactive springs in Auvergne [8].

Diatoms are eukaryotic, photosynthetic microorganisms, accounting for a great part of the carbon dioxide fixation in the aquatic environments. Their dimensions are of the order of a few micrometres and they are uniquely characterized by their frustule, a silicate membrane serving as a skeleton [9] which remains as fossil after their death [10]. In addition to being used as water quality bio-indicators due to their sensitivity and direct response to environmental stresses [11], diatoms are now studied for their response to natural radioactivity in the frame of TIRAMISU project.

The TIRAMISU (biodiversiTy in the RAdioactive Mineral Springs in Auvergne) project is dedicated to the study of the radiological and radiotoxic effects of natural radioactivity on microorganisms living in naturally radioactive mineral springs within ZATU [12]. In the current study, we are focusing on the radiological content of these ecosystems and the simulation of the radiation exposure on diatoms.

In order to characterize the ecosystems, campaigns of sampling were launched including the collection of water, diatoms, sediment, and travertine cores, in 27 locations. The radiological content of the water and sediment samples was analysed using γ -spectroscopy. The study and calculation of the radiation effects on diatoms has been feasible due to Monte Carlo Simulation Codes (MCSC). Especially at the nanoscale, the stochastic nature of the interactions responsible for the DNA damage due to ionizing radiation, make the use of MCSC an essential part of radiological studies [13].

In the current study, an evaluation of the dose rate based on the measured activity values for the biota of freshwater ecosystems is initially performed using the ERICA tool. Microdosimetric calculations follow using GATE, while nanodosimetry is covered by the use of Geant4-DNA.

2 Materials and Methods

2.1 Sampling and Analysis by γ - Spectrometry

Water samples have been collected from 27 mineral springs in Auvergne, Massif Central, France (Figure 1). A High-Purity Germanium (HPGe) well-type detector was used for the spectroscopic analyses at LPC (Clermont Ferrand, France). For the determination of the ^{222}Rn activity, measurements were conducted at least 4 hours post sampling in order to use the 352 keV γ-ray line of ^{214}Pb ($t_{1/2}$ = 27.06 min) under the assumption of radioactive equilibrium.

Sediment samples were also collected from the different sites. The measurements for the determination of ^{226}Ra activity were conducted at least four weeks after the sealing, allowing for the radioactive equilibrium among ^{226}Ra, ^{222}Rn, ^{214}Pb and ^{214}Bi to be reached. These γ- spectroscopic analyses took place at Subatech (Nantes, France) using a coaxial HPGe detector. The data were then used as an input to the simulations described in the next section.

Fig. 1. Map of the sampled mineral springs in Auvergne illustrated according to the radon activity levels in water: green circle – low activity, orange line – medium activity, red star – high activity.

2.2 Simulation

2.2.1 ERICA

ERICA tool v.1.3 [14] has been developed to evaluate dose rates to biota from internal and external distributions of radionuclides. The evaluations are performed using site specific, measured concentrations of radionuclides which are coupled with distribution coefficients for terrestrial and aquatic environments and a database of radiation effects for the model organisms [15].

In the current study, a dose rate assessment was performed for phytoplankton in a freshwater aquatic ecosystem for each individual mineral spring using the measured sediment activity concentrations (in Bq/kg) for ^{226}Ra, ^{238}U and ^{210}Pb. The ^{222}Rn water activity concentrations were excluded due to absence of noble gazes from the radiological factor database of ERICA.

The basic steps followed were the estimation of the total weighted absorbed dose rates (DR in μGy/h) for the phytoplankton in each mineral spring and then, their comparison to the default dose rate screening value for the protection of the ecosystems (SDR= 10 μGy/h) [16]. The risk quotient (RQ) was then assessed as:

$$RQ = \frac{DR}{SDR} \quad (1)$$

For phytoplankton, the concept of risk should be interpreted as the risk of a negative effect on the growth ability with a morbidity endpoint.

In our case, the constraints imposed by ERICA, including the absence of ^{222}Rn in dose rate computations, as well as the need of a complete evaluation of the radiation exposure, highlighted the necessity to use Monte Carlo Simulation Codes.

2.2.2 GATE

Geant4 is an open-source simulation toolkit for the tracking of particles through matter [17] which is well established for modelling and simulating from nanoscale up to macroscale [18]. GATE, an opensource platform based on Geant4 libraries initially dedicated to medical physics applications [19,20], has been employed for the simulation of the absorbed dose to diatoms due to the natural radioactivity present in the mineral springs.

In the current work, we modelled the diatom as a 20 μm radius water sphere surrounded by 100 μm radius water and/or sediment sphere serving as the surrounding environment. When considering sediments, a mixture consisting of a user-defined "dry sediments" material and "G4_WATER" was used, characterized by its porosity values (in %) as described in eq. 2:

$$Porosity = \frac{Volume_{water}}{Volume_{sediments}} * 100 \quad (2)$$

Focusing on the two most abundant radionuclides measured in the water and the sediment of the mineral springs, the alpha spectra of ^{222}Rn and ^{226}Ra were introduced as volumetric, isotropic sources surrounding the diatom and irradiating it either externally and/or internally and the total dose deposited in the diatom (D in Gy) was collected.

For the needs of the Geant4-DNA simulation, information on the type, kinetic energy and direction of the particles entering the diatom were also recorded in a file using the dedicated GATE "PhaseSpace" (PhS) actor.

2.2.3 Geant4-DNA

Geant4-DNA is an open-source Monte Carlo Track Structure (MCTS) code dedicated to micro and nanodosimetry allowing the tracking of particles down to eV energies [21–24] Geant4-DNA allows also the simulation of the indirect DNA damage due to water radiolysis [25] as well as the evaluation of Single and Double Strand Breaks (SSBs and DSBs respectively) due to the direct energy deposition of the ionizing particles through clustering algorithms.

In this study, the assessments of SSBs and DSBs in diatoms is performed using the clustering algorithm DBSCAN [26]. The diatom is represented by a water sphere of 10 μm radius with a spherical nucleus of 0.5 μm. Considering that approximately 10 DNA base pairs (bp) make up one full twist of DNA helix, we are following the modelling suggested by Charlton et al. [27]. The DNA molecules are modelled as nanometric cylinders randomly positioned in the nucleus. Two sizes are considered: 10 nm diameter and 5 nm height representing a nucleosome and 2 nm diameter and 2 nm height representing 10 bp. The source simulated is the PhS file containing only the alpha particles which was collected from the GATE simulation described in section 2.2.2. The specific energy (z in Gy), corresponding to the dose deposited in the cylinders, is then collected.

According to the DBSCAN clustering algorithm, the formation of SSBs is considered a function of the energy deposition following a probability distribution function. For deposited energy (e_{dep}) < 5 eV the damage probability is considered zero while it increases linearly up to 37.5 eV. For $e_{dep} \geq 37.5$ eV the algorithm considers that all the events can cause SSBs. The minimum number of SSBs to form a DSB is set to 2 within a radius of 3.3 nm, representing roughly the distance between 10 DNA base pairs.

3 Results and Discussion

3.1 γ- Spectrometry

The ^{222}Rn concentration activity (A_{Rn} in Bq/L) values in the waters of the mineral springs range between 1.3 and 4000 Bq/L. The mineral springs were classified according to the values of ^{222}Rn activities in three categories of low ($A_{Rn} < 100$ Bq/L), medium ($100 \leq A_{Rn} <$ 1000 Bq/L) and high activity ($A_{Rn} \geq 1000$ Bq/L) levels. In overall, 65% of the sampled locations belong to the low activity level class while 19% belong to the high activity level class. Among all the mineral springs, La Montagne is presenting the highest values of ^{222}Rn activity concentration in the water, with a maximum of 4000 Bq/L so far.

According to the analyses, ^{210}Pb is present in the sediments in 80% of the mineral springs with a mean specific activity value of 1.4 Bq/g. ^{238}U and ^{226}Ra are present in the sediments of all the mineral springs with specific activities ranging between 7.6 10^{-3} and 0.5 Bq/g, and 0.028 to 52 Bq/g respectively.

3.2 Theoretical simulations

Using the activity values measured by γ-spectrometry, the dose rate assessment for the phytoplankton in the mineral springs indicated that the contribution of ^{238}U and ^{210}Pb to the dose rate is very small. On the other hand, it was found that the dose rate to phytoplankton exceeded the threshold limit (10 μGy/h) in 71% of the sampled spots due to the presence of ^{226}Ra in the sediments. In Figure 2, the bar plot shows the total estimated dose rate (in μGy/h) corresponding to the ^{226}Ra specific activity (Bq/kg) of each spring. The solid line represents the respective values of the ^{226}Ra specific activity, while the dashed line indicates the dose rate threshold limit (10 μGy/h) for ERICA risk assessment in the ecosystems.

The dose rate values range from 0.17 μGy/h for Poix up to 125 μGy/h for Daguillon. A clear correlation between the ^{226}Ra specific activity and the dose rate is also observed.

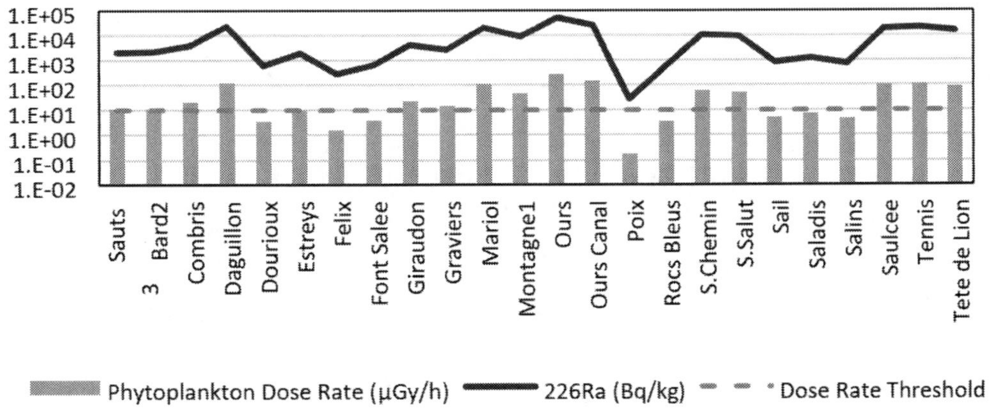

Fig. 2. Radium specific activity in sediments and dose rate for phytoplankton in each spring.

The dose rate (in µGy/h) received by the diatoms as a function of the porosity of the sediments is presented in Figure 3. The specific activities and activity concentrations here represent mean real values that have been measured from the collected samples. The porosity values at the two edges of the x-axis correspond to a totally dry sediments composition (0% of water) in the case of 0% porosity, while 100% porosity refers to a composition of 100% water.

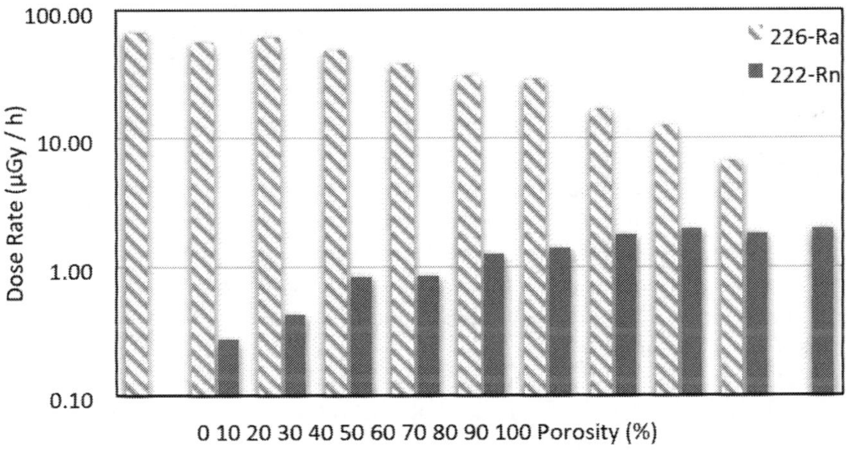

Fig. 3. Dose rate to the diatoms as a function of porosity in the sediments scaled for 30 Bq/gr of ^{226}Ra and 1000 Bq/L of ^{222}Rn.

The observed trend is that the dose rate received by the diatoms decreases as a function of the porosity. That indicates the higher contribution of the ^{226}Ra (present in the sediments) to the dose received by the microorganisms living in mineral springs compared to the dose imposed by the presence of ^{222}Rn in the water.

The impact of the internal and/or external exposure to the dose rate received by the diatoms was also evaluated. An increase of the dose rate is observed in the case of the combined internal/external exposure for both the mean measured values of ^{222}Rn in the water (1000 Bq/L) and ^{226}Ra the sediments (30 Bq/g). The result can be interpreted by the fact that both radionuclides are alpha emitters of relatively small energies (5.5 MeV and 4.8 MeV respectively) which deposit their energy in a few tens of micrometers. In water, their energy range is well known [28], while in the case of the sediments the simulation offers a first evaluation of the range of the alpha particles in this specific medium.

The nanodosimetric simulations for the case of a nucleosome surrounded by water with 1000 Bq/L ^{222}Rn activity concentration assessed a specific energy rate of 0.023 μGy/h. The respective simulation for 10 base pairs, resulted in a specific energy rate of 8.4 10^{-4} μGy/h. The observed 2-orders-of-magnitude difference can be attributed to the different sizes of the modelled molecules.

For the case of the simulation of the direct DNA damage using the clustering algorithm, 131 SSBs per particle and 57 DSBs per particle were predicted. These preliminary results can only offer an order of magnitude for the DNA damages expected in the diatom.

4 Conclusions

In this study, we simulated the effects of natural radioactivity on microorganisms living in radioactive mineral springs. This work contributes to the part of the understanding and the evaluation of the doses received to diatoms in the mineral waters in Auvergne region in France.

In total, 27 mineral springs have been sampled and analyzed. The highest ^{222}Rn activity concentration measured in water was 4000 Bq/L, while the highest ^{226}Ra specific activity in the sediments was 52 Bq/g.

Considering the measured activity values, a dose rate estimate for aquatic freshwater ecosystems has been performed using the ERICA tool. The results indicate dose rates above ERICA risk threshold imposed to the phytoplankton in 71% of the mineral springs observed as a result of the high ^{226}Ra concentration in the sediments.

Microdosimetric simulations have been conducted using the GATE open-source Monte Carlo platfom. The results have shown among others that the highest contribution to the dose received by the diatoms is due to the alpha particles present in the natural decay series. From this study, it was also evaluated that the most important contributor to the dose is the ^{226}Ra which is present in the sediments of the mineral spring ecosystem. The investigation of the case of only external or a combination of internal and external radiation exposure has highlighted the necessity of dedicated studies. These studies are currently investigating the capacity of the radionuclides to penetrate the silicate frustule of the diatoms.

The simulated impact of the sediment composition on the dose received by the diatoms has also provoked the launch of measurements for the complete characterization of the chemical contents in the sediments.

Nanodosimetric simulations using Geant4-DNA have given an order of magnitude for the specific energy rate on the DNA of the diatoms due to the presence of ^{222}Rn in the water, while simulations are currently performed ^{226}Ra in the sediments, too. The values range between 0.023 μGy/h and 8.4 10^{-4} μGy/h as a function of the size of the modelled molecule for a ^{222}Rn source isotropically distributed in water.

SSBs and DSBs were calculated for ^{222}Rn in water, showing a trend of two times more single than double strand breaks. These preliminary results will be supplemented by the respective ones for ^{226}Ra in sediments. In order to be used for a complete DNA damage evaluation, these results need to be consolidated with additional indirect damage caused by chemical species after radiolysis, a work which is currently in progress.

Finally, the response of the diatoms to external alpha and X-ray beams is currently being investigated in an effort to couple the predicted DNA damage results stemming from the simulation to observables from experiments.

The studies have showed so far that there is a great potential for the evaluation of the link between natural radioactivity and the diatom teratogenic forms observed in natural radioactive mineral springs.

This work was funded by CNRS in the frame of "Prime80 CNRS projects".

References

1. M. E. Kitto *et al.*, J. Environ. Radioact. **80**, 327 (2005).
2. S. A. Saqan, M. K. Kullab, and A. M. Ismail, J. Environ. Radioact. **52**, 99 (2001).
3. NNDC, NuDat2, (https://www.nndc.bnl.gov/nudat3/nudat2.jsp).
4. N. Saclier *et al.*, Elife **9**, 1 (2020).
5. G. Ielsch *et al.*, J. Environ. Radioact. **166**, 210 (2017).
6. A. Beauger *et al.*, BIOM **1**, 21 (2020).
7. G. G. Lai *et al.*, PeerJ **7**, (2019).
8. F. Millan *et al.*, Bot. Lett. **167**, 95 (2020).
9. N. Kröger and N. Poulsen, Annu. Rev. Genet. **42**, 83 (2008).
10. D. M. Harwood, V. Nikolaev, and D. M. Winter, Paleolontogical Soc. Pap. **13**, 33 (2007).
11. R. J. Stevenson, Y. Pan, and H. van Dam, Diatoms Appl. Environ. Earth Sci. Second Ed. 57 (2010).
12. ZATU, (https://deims.org/915cb2eb-ed08-4b02-9d16-1f4d688e912b).
13. K. P. Chatzipapas *et al.*, Cancers (Basel). **12**, 1 (2020).
14. J. E. Brown *et al.*, J. Environ. Radioact. **99**, 1371 (2008).
15. D. Copplestone, J. Hingston, and A. Real, J. Environ. Radioact. **99**, 1456 (2008).
16. J. Garnier-Laplace *et al.*, J. Environ. Radioact. **99**, 1474 (2008).
17. S. Agostinelli *et al.*, Nucl. Instruments Methods Phys. Res. Sect. A Accel. Spectrometers, Detect. Assoc. Equip. **506**, 250 (2003).
18. S. Incerti *et al.*, Phys. Medica **32**, 1187 (2016).
19. A. Bongrand *et al.*, Cancers (Basel). **13**, (2021).
20. P. Papadimitroulas, Phys. Medica **41**, 136 (2017).
21. S. Incerti *et al.*, Med. Phys. **37**, 4692 (2010).
22. M. A. Bernal *et al.*, Phys. Medica **31**, 861 (2015).
23. S. Incerti *et al.*, Med. Phys. **45**, e722 (2018).
24. S. Incerti *et al.*, Int. J. Model. Simulation, Sci. Comput. **1**, 157 (2010).
25. D. Peukert *et al.*, Med. Phys. **46**, 983 (2019).

26. Z. Francis, C. Villagrasa, and I. Clairand, Comput. Methods Programs Biomed. **101**, 265 (2011).

27. D. E. Charlton, H. Nikjoo, and J. L. Humm, Int. J. Radiat. Biol. **56**, 1 (1989).

28. NIST, ASTAR, (https://physics.nist.gov/PhysRefData/Star/Text/ASTAR.html).

A Review of Novel Instrumentation for Matrix Independent Ultratrace Analysis of Radionuclides using Collinear Resonance Ionisation Spectroscopy (CRIS)

Giles Edwards[1, 2], *Holly* Perrett[1, 2], *Matthew* Duggan[1, 2], *Jordan* Reilly[1, 2] and *Kieran* Flanagan[1, 2*]

[1]Department of Physics, The University of Manchester, Manchester, United Kingdom
[2]Photon Science Institute, The University of Manchester, Manchester, United Kingdom

Abstract. The CRIS setup at CERN-ISOLDE is a high resolution laser spectroscopy experiment that is used to characterise exotic radionuclei with very low production rates. Acquisition of hyperfine spectra enable nuclear spins and precise values of nuclear magnetic dipole and electric quadrupole moments to be determined. The high sensitivity of the CRIS technique combined with the high selectivity make it an ideal candidate for quantitation of analytes in complex matrices. This review article reports on novel instrumentation under development at the University of Manchester, utilising the CRIS technique for targeted matrix independent ultra-trace analysis.

1 CRIS Experiment at CERN-ISOLDE

The CRIS experiment at CERN-ISOLDE was first reported in 2013 and uses laser spectroscopy to measure fundamental nuclear properties through acquisition of hyperfine spectra, from which nuclear observables can be extracted [1]. In this way, nuclear moments, relative charge radii and nuclear spins can be determined without relying on assumptions from theoretical models when comparing nuclear properties of isotopes of the same element. It is possible to observe how the shape of the nucleus changes with the addition or subtraction of neutrons.

The CRIS technique combines the high resolution of collinear laser spectroscopy (CLS) with the high sensitivity and efficiency of resonant ionisation spectroscopy (RIS).

The current configuration of the CRIS setup at ISOLDE allows exotic radionuclei to be studied, as well as stable isotopes using a laser ablation source [2]. Exotic radionuclei are produced by impact of the CERN PSBooster proton beam (1.4 GeV) onto the On-Line Mass Separator (ISOLDE) UC_x target. Radionuclei diffuse out of the target, and the resulting beam is mass separated to select the isotope of interest, collisionally cooled and bunched using a gas-filled radiofrequency linear Paul trap and accelerated to 30-50 keV, ready for delivery to the CRIS beam line Accelerating the ions increases the selectivity of the technique compared with in-source RIS by introducing a kinematic frequency shift of spectral lines for different isotopes.

* Corresponding author: kieran.flanagan-2@manchester.ac.uk

The bunched ion beam is neutralized in flight using a heated charge exchange cell loaded with alkali metal vapor, after which any remaining charged species are deflected off axis.

The resulting atom bunches are overlapped with one or more laser beams and synchronised with the laser duty cycle in order to maximise efficiency. The excitation laser frequency is scanned across the electronic transition under investigation, exciting the atoms, which may then be ionized using an additional laser or an electric field. The ions are deflected off axis and detected by an ETP MagneTOF detector. The laser-atom interaction region is maintained at $\sim 10^{-9}$ mbar in order to minimise interference from collisional activated ionization. Recent additions to the beamline include a set of field ionization grids post interaction region, for the ionization of Rydberg atoms [5].

The CRIS experiment at CERN-ISOLDE has successfully completed campaigns on elements of various exotic radionuclei including francium [1], copper [4], radium [6], indium [7], and potassium [8, 9]. More recently isotope shifts of radium monofluoride molecules have been determined [10].

CRIS has performed measurements with short lived nuclei and exotic beam production rates low as 10 atoms per second. Moreover, reduction of collisional background due to interference species by 7 orders of magnitude has been demonstrated [4].

The high sensitivity and selectivity of the CRIS technique allow for its application to both fundamental nuclear physics studies and challenging analytical chemistry based assays.

2 CRIS Experiment at Manchester

Quantitative, label free elemental analysis is often compromised by sample matrix effects. The response of the analyte with most techniques is either enhanced or suppressed by the chemical environment that makes up the remainder of the bulk material (sample matrix). Various strategies exist that endeavor to mitigate these matrix effects. These can include additional sample preparation procedures that can remove matrix interference. Alternate methods of matrix independent quantitation often involve ionization and subsequent quantitation of the analyte in the gas phase. Matrix interference is often significantly reduced when ionising neutrals in the gas phase.

With kinematic isotope separation as well as mass separation with magnets, the capability to measure atomic transitions with linewidths of 25 MHz [11], and demonstrated sensitivities of 10s of atoms per second [4], the CRIS technique enables a specific regime for targeted analysis with matrix independent ultra-trace limits of detection. A CRIS beamline is under development in the Photon Science Institute at the University of Manchester for this purpose. The configuration of the instrument is similar to that at CERN-ISOLDE, incorporating mass selection of the analyte followed by resonant excitation and ionisation, which facilitates additional selectivity above and beyond what is possible with nominal mass separation. Multiple experiments will be carried out on the CRIS instrument in Manchester, which are described in the following sections. Depending on the application, the instrumentation may employ either electron ionisation (EI), electron cyclotron resonance (ECR) or inductively coupled plasma (ICP) as the ion source.

The ion beam is accelerated to 3 - 5 keV and passed through a series of steering and focusing optics. In order to make the experimental set up as compact as possible, the

Manchester CRIS beamline is on 2 levels. Ions are produced at one end of the lower beamline and travel through a set of electrodes to steer and focus the beam onto the top level. A linear Paul trap cooler buncher based on previous work [13, 14] (figure 1 RFQcb), currently under development, will be installed to produce a bunched beam compatible with the duty cycle of the laser system. The ion beam is focused through a charge exchange cell where it is neutralised in flight. Ions that are not neutralised are deflected out of the beam. The charge exchange cell is heated to 513 K and uses sodium vapour as the charge exchange media.

The excitation laser or lasers are steered into the beamline on the top level, overlapping with the fast atom beam. Atoms neutralised into the chosen spectroscopic state will be resonantly excited to a state from which they will be ionised. The ionisation step will be provided by one of two methods. For the first method, the 532 nm output of a Nd:YAG laser will be steered through the upper beamline to ionise the excited atoms. For the second method, potentials will be applied to 3 wire grids, mounted on printed circuit boards and spaced 4 mm apart, which are positioned in the path of the neutral beam. The resulting electric field will ionise Rydberg atoms passing through the region by field ionisation [5].

An electron multiplier detector is positioned offaxis. A final pair of deflectors steer ions into the electron multiplier, where they will be counted. The resulting signal will be gated on the laser pulse timing. Later iterations of the CRIS beamline will utilise a multicollector magnetic sector for mass separation.

2.1 Detection of Krypton-85

^{85}Kr is a radioactive noble gas present in the atmosphere. Although a small amount of ^{85}Kr is produced naturally through the interaction of cosmic neutrons with stable isotopic neighbour ^{84}Kr, the majority originates from anthropogenic sources. As a fission product, ^{85}Kr is formed in nuclear reactors. Unless the fuel cladding is damaged, the ^{85}Kr formed is retained in spent fuel rods, which are usually kept in storage for a minimum of 6 months after use in order to reduce their activity. An increase in the atmospheric concentration of ^{85}Kr in the vicinity of a storage facility can indicate a leak in fuel containers. When spent fuel rods are reprocessed, most of the ^{85}Kr is released into the atmosphere, and is difficult to contain. Considering these implications, accurate ^{85}Kr measurement also has applications in nuclear security, as a spike in local atmospheric concentration could indicate the presence of an undeclared reprocessing facility [12].

^{85}Kr is present in the atmosphere with an isotopic abundance of 10^{-12} and atmospheric concentration of 10^{-18}. It is therefore necessary that ^{85}Kr detection requires both high sensitivity and mass resolution (abundance sensitivity) to be able to resolve the ^{85}Kr signal over the much larger ^{84}Kr component, as well as distinguishing it from any isobaric interference. After initial proof of principle studies the beamline will be optimised for the measurement of ^{85}Kr/^{84}Kr isotopic ratios.

For initial studies, CRIS will be performed on argon atoms. In the current configuration, argon gas is injected into an EI ion source at a rate of 1 mL min^{-1}, where singly-charged argon ions are produced and extracted through 4 electrostatic lenses. Currently, argon beams of around 0.25 μA are measured immediately after extraction. Simulations based on electron-capture cross section calculations [15] have demonstrated that after neutralisation by sodium vapour, the metastable $4s^2{}_{[3/2]}0$ state in argon will be

42.6% populated. From this state, argon atoms can be excited to high-lying Rydberg states through 1-photon absorption of 300 nm light [16].

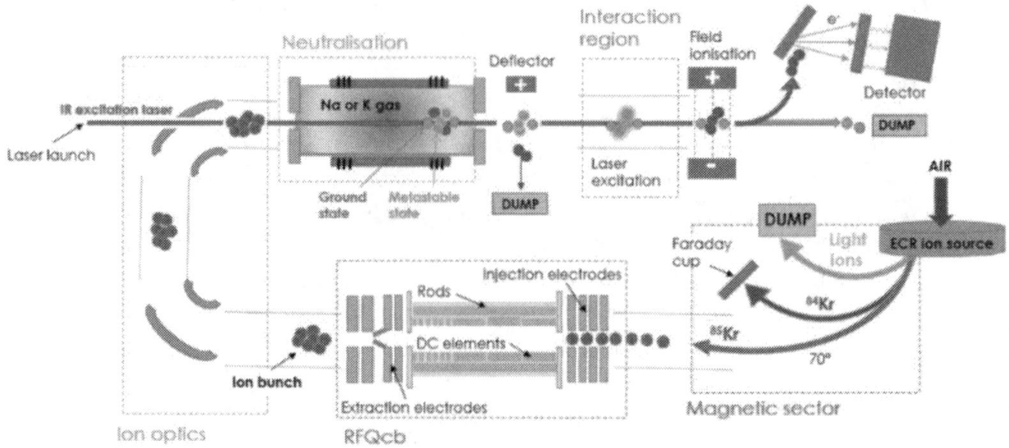

Fig. 1. Manchester CRIS setup with ECR gas inlet configured for ^{85}Kr analysis with multi-collector magnetic sector

2.2 Radiocarbon Dating

Accelerator Mass Spectrometry (AMS) is the industry standard method for radiocarbon dating applications. When used for radiometric dating, AMS employs a multi-collector detector array for isotope ratio mass spectrometry. The most common method of sample introduction involves complete combustion of the sample in pure oxygen to evolve carbon dioxide gas; this is then graphitised and loaded into a metal target alongside reference materials used as external standards for quantification. A primary caesium ion gun is directed at the samples to produce secondary negative carbon ions which are then accelerated to the stripper in the centre of a tandem accelerator. The stripper removes the negative charge and reduces the negative carbon ions to a triply charge positive carbon ion which are then accelerated out of the tandem accelerator for subsequent mass selection. The carbon ions are resolved according to their *m/z* ratio in order to resolve the ^{12}C, ^{13}C and cosmogenic ^{14}C isotopes. The relative isotopic abundances may be used to calculate the date at which biological sample deceased. when using relatively small accelerating potentials on AMS instruments, it is not always possible to perform radiometric dating of larger cosmogenic radionuclides such as ^{26}Al or ^{36}Cl.

The CRIS beamline in Manchester will be optimised for the detection of the low abundance ^{14}C isotope, in order to perform radiocarbon dating. Samples submitted for carbon dating analysis are first combusted in pure oxygen using a flash gas chromatograph. The combustion gases are then separated on column, reducing the nitrogen background by a factor of approximately 10^6. Further suppression and removal of nitrogen gas interference is achieved by means of a cryogenic concentrator. The eluting carbon dioxide combustion fraction is diverted post column into a cryogenic trap cooled to 77 K using a 10-port Valco switching valve. The carbon dioxide elutes in a helium carrier gas with a flow rate of 120 mLmin^{-1} .Once trapped the solid carbon dioxide is ready for analysis. The 10-port valve switches an additional helium flow set at 1 $mLmin^{-1}$ into the trap which is then heated to room temperature. The carbon dioxide sublimes and is swept into an ECR ion source to be ionised and fragmented into singly charged positive carbon ions, which will ultimately be delivered to the main beamline. The CRIS setup will be operated in the same way described earlier for the ^{85}Kr analysis, employing the same multi-collector magnetic sector mass analyser used for the ^{85}Kr application depicted in figure 1. The magnet would be tuned for the separation of the carbon isotopes, and the lasers will be tuned for resonant excitation of ^{14}C.

The application of CRIS to radiometric dating applications offers advantages when compared to AMS techniques. One advantage is that the lasers may be tuned to be on resonance with almost any element from across the periodic table, which would allow for radiometric dating of ^{26}Al and ^{36}Cl. Additionally, the use of an ECR ion source and direct analysis of the carbon dioxide from combusted samples negates the need for a graphitisation step. This reduces this significant bottle neck often experienced when using AMS techniques, as the graphitisation process itself introduces additional experimental errors associated with sample preparation. The capacity of the ECR ion source to produce beams of > 1 mA will also significantly speed up the radiometric dating process.

2.3 Hyphenation of CRIS with ICP-MS

The hyphenation of the CRIS technique to industry standard ultra-trace analytical techniques, such as inductively coupled plasma mass spectrometry (ICPMS) offers many advantages including the ease of integration with existing workflows. CRIS has the potential to offer additional sensitivity and selectivity for challenging applications that require limits of currently not available even with state-of-the-art tandem mass analyser based ICP-MS/MS instrumentation.

Many applications of radioisotope analysis in the field of nuclear decommissioning require limits of detection not currently achievable with the latest instrumentation, but would be made possible with ICP-MS-CRIS. In particular, this novel technique will be used for the quantification of ^{90}Sr, which is a high yield fission product and is continuously created in civil nuclear power plants. The long half-life, environmental mobility and ability of ^{90}Sr to concentrate within the food chain make this an important radionuclide for nuclear decommissioning and waste characterisation.

In many applications, such as processing plant waste sentencing, ^{90}Sr levels are inferred by determining a gamma spectroscopy "fingerprint" for a substrate and then measuring concurrently and reliably produced ^{137}Cs. However, the differing partitioning of ^{90}Sr and ^{137}Cs in the environment, particularly in soil and groundwater, invalidates any 'fingerprinting' and both isotopes must be measured independently. The existing method for ^{90}Sr quantification comprises of Cherenkov counting of the daughter isotope ^{90}Y. This requires complete chemical separation of ^{90}Sr and a >20 day grow in period of the ^{90}Y daughter isotope which constitutes a major bottleneck in the analysis of ^{90}Sr. At the present time only the Cherenkov counting method is accredited and sensitive enough to detect ^{90}Sr present in environmental samples which are set with an upper limit of 0.5 BqL^{-1} in groundwater and 8 $Bqkg^{-1}$ in soil in the United Kingdom.

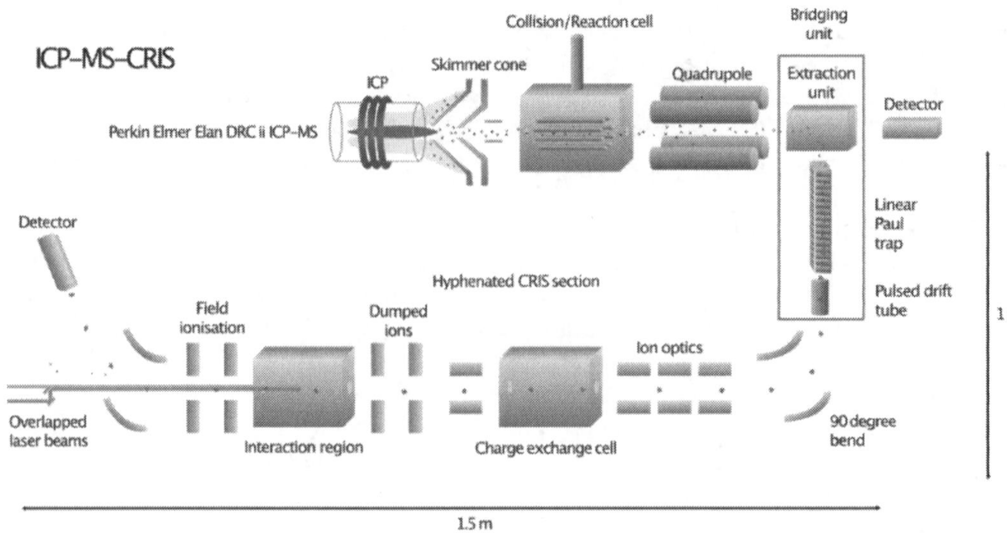

Fig. 2. Manchester CRIS setup with hyphenation to ICP-MS instrumentation for ICP-MS-CRIS

As the nuclear decommissioning programme accelerates, the ability to characterise larger volumes and sample numbers requires a cost-effective solution. Additionally, the practical requirements of site characterisation, remediation and verification/validation demand a quantification technique with a rapid capability, ideally with a turnaround time of a few hours. ICP-MS-CRIS has the capacity to fulfil these requirements.

In addition to ^{90}Sr, there also exists a range of short-lived isotopes which are important within the decommissioning process that also require multi-stage chemical separation before they can be measured.

The hyphenation of ICP-MS with CRIS will be achieved via a bridging unit between an ICP-MS instrument and a CRIS beamline. The bridging unit will consist of an ion extraction unit, linear Paul trap and ion optics. The ion extraction unit is comprised of two ion carpets and three repeller plates that divert ions leaving the ICP-MS quadrupole mass analyser through a 90° bend into the linear Paul trap. The ion carpet is made up of 24 concentric circular rings originating from a 2 mm entrance aperture, and a series of rectangular repeller plates shown in figure 3. The ion extraction unit will be located in a 46 x 123 mm region between the ICP-MS quadrupole and the detector.

The carpet operates with a DC potential gradient, known as the drift potential, applied across the ring electrodes to guide ions towards the aperture and into the linear Paul trap. When operating in the 10^{-2} mbar range, the buffer gas reduces the ion kinematics.

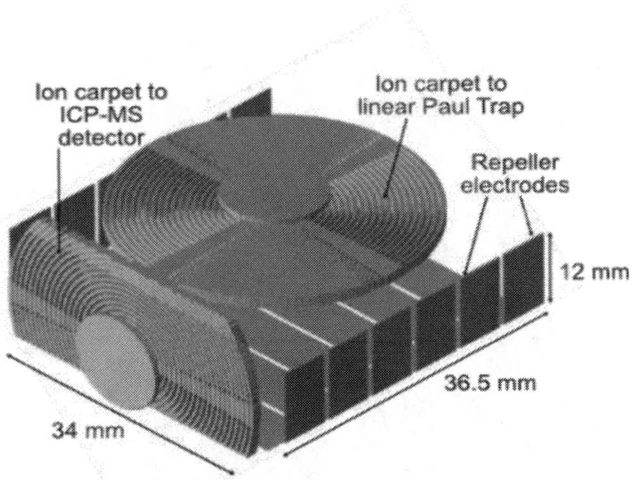

Fig. 3. SIMION 3D representation of the ion extraction unit. A series of repeller plates perpendicular and opposite to two ion carpets.

An RF potential applied to a set of electrodes causes the ions to hover and oscillate over the carpet while a DC potential directs the ions towards the aperture.

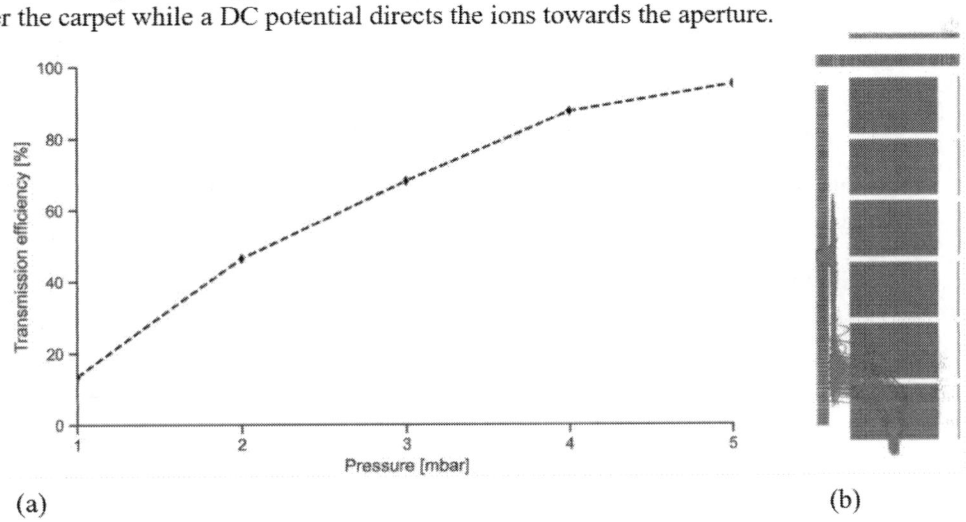

Fig. 4. (a) Simulated ion transmission in the mbar range (b) Example of ion motion under applied DC and RF potentials at 5 mbar.

SIMON simulations, using *m/z* 90 to represent ^{90}Sr, indicate a near 100% ion transmission efficiency in the mbar region; as exemplified in Figure 4. At pressures below 10^{-2} mbar, the DC potential begins to dominate as the RF potential diminishes and becomes negligible due to the reduced buffer gas pressure. As the linear Paul trap has an operating pressure of 10^{-2} mbar and is mounted directly onto the extraction unit, the operating pressure of the extraction unit will be between 10^{-6} mbar (ICP-MS operating pressure) and 10^{-2} mbar.

Simulations at the upper limit of operating pressure, shown in Figure 6, indicate an ion transmission efficiency of around 40-45%, however, by optimising the extraction unit geometry, the anticipated transmission efficiency will likely increase.

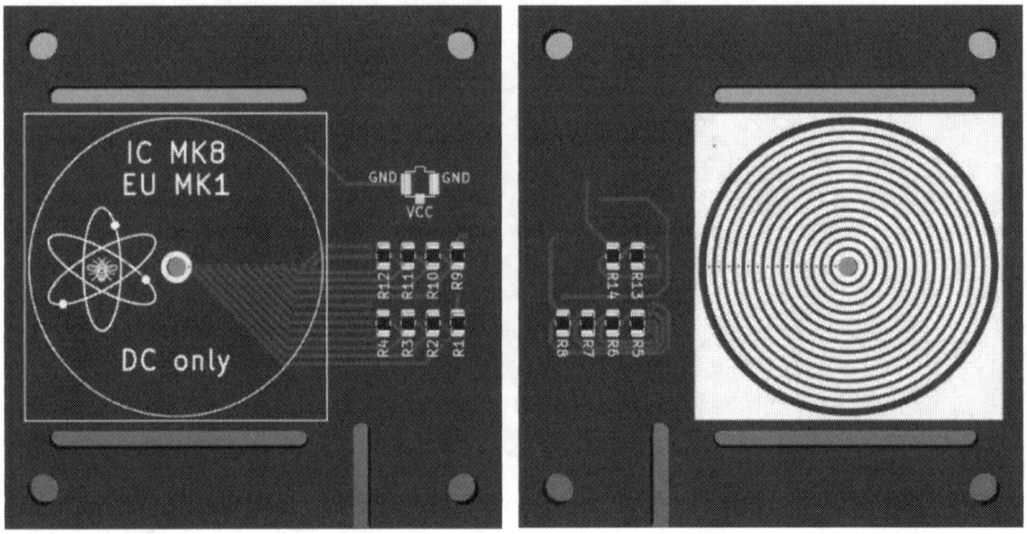

(a) (b)

Fig. 5. 3D rendering of the top (a) and bottom (b) panels of the ion carpet which extracts the ions that traversed through the ICP-MS quadruple.

The ion extraction unit will be constructed using printed circuit boards.. The extraction unit will be tuned to transmit ions of various *m/z* values. The top and bottom panels of the extracting ion carpet is shown in Figure 5 (a) and (b) with a resistor chain determining the potential on each ring electrode. The planar linear Paul trap along with the ion extraction unit, will be investigated to determine its suitability for the hyphenation of ICP-MS with CRIS.

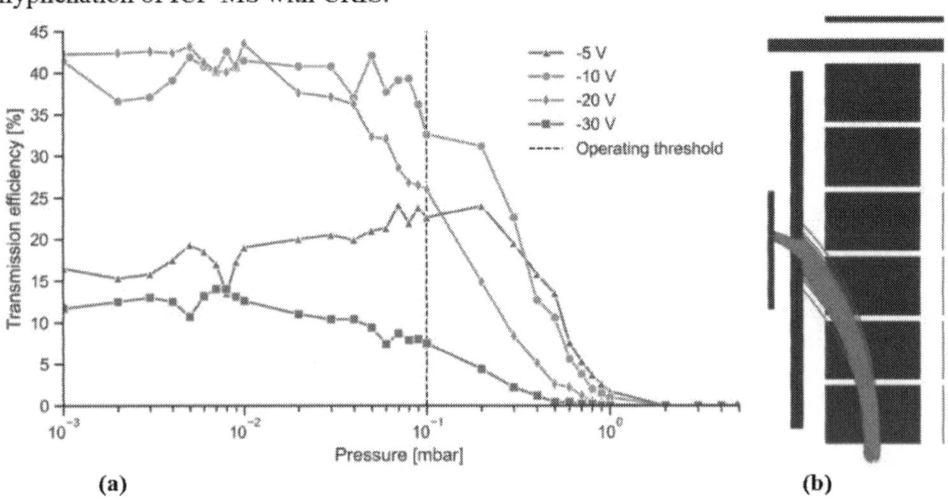

(a) **(b)**

Fig. 6. (a) Simulated ion transmission in the mbar range (b) Example of ion motion with only an applied DC potential at 1 x 10^{-2} mbar.

References

1. K. T. Flanagan, K. M. Lynch, J. Billowes, M. L. Bissell, I. Budincevic, T. E. Cocolios, R. P. de Groote, S. De Schepper, V. N. Fedosseev, S. Franchoo, R. F. Garcia Ruiz, H. Heylen, B. ,A. Marsh, G. Neyens, T. J. Procter, R. E. Rossel, S. Rothe, I. Strashnov, H. H. Stroke, and K. D. A. Wendt. Phys. Rev. Lett. **111**, 212501 (2013)
2. R. F. Garcia Ruiz, A. R. Vernon, C. L. Binnersley, B. K. Sahoo, M. Bissell, J. Billowes, T. E. Cocolios, W. Gins, R. P. de Groote, K. T. Flanagan, A. Koszorus, K. M. Lynch, G. Neyens, C. M. Ricketts, K. D. A. Wendt, S. G. Wilkins, and X.F. Yang. Phys. Rev. X **8**, 041005 (2018)
3. S. Rothe, T. D. Goodacre, D. Fedorov, V. Fedosseev, B. Marsh, P. Molkanov, R. Rossel, M. Seliverstov, M. Veinhard, and K. Wendt, Nucl. Instrum. Methods Phys. Res., Sect. B **376**, 91, (2016)
4. R. P. de Groote, J. Billowes, C. L. Binnersley, M. L. Bissell, T. E. Cocolios, T. Day Goodacre, G. J. Farooq-Smith, D. V. Fedorov, K. T. Flanagan, S. Franchoo, R. F. Garcia Ruiz, Á. Koszorús, K. M. Lynch, G. Neyens, F. Nowacki, T. Otsuka, S. Rothe, H. H. Stroke, Y. Tsunoda, A. R. Vernon, K. D. A. Wendt, S. G. Wilkins, Z. Y. Xu, and X. F. Yang. Phys. Rev. C **96**, 041302(R) (2017)
5. A. R. Vernon, C. M. Ricketts, J. Billowes, T., E. Cocolios, B. S. Cooper, K. T. Flanagan, R. Garcia Ruiz, F. P. Gustafsson, G. Neyens, H. A. Perrett, B. K. Sahoo, Q. Wang, F. J. Waso, and X. F. Yang. Sci. Rep. **10**, 12306 (2020)
6. K. M. Lynch, S. G. Wilkins, J. Billowes, C. L. Binnersley, M. L. Bissell, K. Chrysalidis, T. E. Cocolios, T. Day Goodacre, R. P. de Groote, G. J. Farooq-Smith, D. V. Fedorov, V. N. Fedosseev, K. T. Flanagan, S. Franchoo, R. F. Garcia Ruiz, W. Gins, R. Heinke, Á. Koszorús, B. A. Marsh, P. L. Molkanov, P. Naubereit, G. Neyens, C. M. Ricketts, S. Rothe, C. Seiffert, M. D. Seliverstov, H. H. Stroke, D. Studer, A. R. Vernon, K. D. A. Wendt, and X. F. Yang. Phys. Rev. C **97**, 024309 (2018)
7. B. K. Sahoo, A. R. Vernon, R. F. Garcia Ruiz, C. L. Binnersley, J. Billowes, M. L. Bissell, T. E. Cocolios, G. J. Farooq-Smith, K. T. Flanagan, W. Gins, R. P. de Groote, Á. Koszorús, G. Neyens, K. M. Lynch, F. Parnefjord-Gustafsson, C. M. Ricketts, K. D. A. Wendt, S. G. Wilkins, and X. F. Yang. New J. Phys. **22**(1), 012001 (2020)
8. Á. Koszorús, X. F. Yang, J. Billowes, C. L. Binnersley, M. L. Bissell, T. E. Cocolios, G. J. Farooq-Smith, R. P. de Groote, K. T. Flanagan, S. Franchoo, R. F. Garcia Ruiz, S. Geldhof, W. Gins, A. Kanellakopoulos, K. M. Lynch, G. Neyens, H. H. Stroke, A. R. Vernon, K. D. A. Wendt, and S. G. Wilkins. Phys. Rev. C **100**, 034304 (2019)
9. Á. Koszorús, X. F. Yang, W. G. Jiang, S. J. Novario, S. W. Bai, J. Billowes, C. L. Binnersley, M. L. Bissell, T. E. Cocolios, B. S. Cooper, R. P. de Groote, A. Ekström, K. T. Flanagan, C. Forssén, S. Franchoo, R. F. Garcia Ruiz, F. P. Gustafsson, G. Hagen, G. R. Jansen, A. Kanellakopoulos, M. Kortelainen, W. Nazarewicz, G. Neyens, T. Papenbrock, P.G. Reinhard, C. M. Ricketts, B. K. Sahoo, A. R. Vernon, and S. G. Wilkins Nat. Phys. , (2021)
10. S. M. Udrescu, A. J. Brinson, R. F. Garcia Ruiz, K. Gaul, R. Berger, J. Billowes, C. L. Binnersley, M. L. Bissell, A. A. Breier, K. Chrysalidis, T. E. Cocolios, B. S. Cooper, K. T. Flanagan, T. F. Giesen, R. P. de Groote, S. Franchoo, F. P. Gustafsson, T. A. Isaev, Á. Koszorús, G. Neyens, H. A. Perrett, C. M. Ricketts, S. Rothe, A. R. Vernon, K. D. A. Wendt, F. Wienholtz, S. G. Wilkins, and X. F. Yang Phys. Rev. Lett. , (2021)
11. S. G. Wilkins, K. M. Lynch, J. Billowes, C. L. Binnersley, M. L. Bissell, T. E. Cocolios, T. Day Goodacre, R. P. de Groote, G. J. Farooq-Smith, K. T. Flanagan, S. Franchoo, R. F. Garcia

Ruiz, W. Gins, H. Heylen, Á. Koszorús, Neyens, H. H. Stroke, A. R. Vernon, K. D. A. Wendt, and X. F. Yang. Phys. Rev. C **96**, 034317 (2017)

12. Kalinowski, M. B.(1994). Proceedings of the workshop on the state of measurement technologies for identification of nuclear signatures. EUR17312, pp 67-72
13. B. S. Cooper, H. A.Perrett, C. M. Ricketts, C. Read, G. Edwards, K. T. Flanagan, J. Billowes, C. L. Binnersley, M. L. Bissell, T. E. Cocolios, R. P. de Groote, G. J. Farooq-Smith, R. F. Garcia Ruiz, W. Gins, Á. Koszorús, G. Neyens, F. P-Gustafsson, H. H. Stroke, A. R. Vernon, K. D. A. Wendt, S. G. Wilkins, and X. F. Yang Hyperfine Interact. **240**, 52 (2019)
14. C. M. Ricketts, B. S. Cooper, G. Edwards, H. A. Perrett, J. Billowes, C. L. Binnersley, T. E. Cocolios, K. T. Flanagan, R. F. Garcia Ruiz, R. P. de Groote, F. P. Gustafsson, Á. Koszorús, G. Neyens, A. R. Vernon, and X. F. Yang. Nucl. Instrum. Methods Phys. Res. B **463**, 375-377 (2020)
15. A. R. Vernon, J. Billowes, C. L. Binnersley, M. Bissell, T. E. Cocolios, G. J. Farooq-Smith, K. T. Flanagan, R. F. Garcia Ruiz, W. Gins, R. P. de Groote, A. Koszorus, K. M. Lynch, G. Neyens, C. M. Ricketts, K. D. A. Wendt, S. G. Wilkins, and X.F. Yang. Spectrochim. Acta. B **153**, 61-83 (2019)
16. M. Pellarin, Phys B-At. Mol. Opt. **21** (23), pp.3833-3849 (1988)

X-ray spectroscopy study on Thai amulet of Phra Kru Nadune

Krit Won-in[1], *Chatdanai* Boonruang[2,3*], *Udomrat* Tippawan[2], *Chome* Thongleurm[4], and *Pisutti* Dararutana[5]

[1]Department of Earth Sciences, Faculty of Science, Kasetsart University, Bangkok 10900 Thailand
[2]Department of Physics and Materials Science, Faculty of Science, Chiang Mai University, Chiang Mai 50200 Thailand
[3]Center of Excellence in Materials Science and Technology, Chiang Mai University, Chiang Mai 50200 Thailand
[4]Science and Technology Research Institute, Chiang Mai University, Chiang Mai 50200 Thailand
[5]Independent Researcher, Retired Army Officer at the Royal Thai Army, Lopburi 15000 Thailand

Abstract. Thai Amulets are found in many forms in Thailand. They have been made by many different ethnic groups with different traditions. Phra Kru Nadune is one of the important amulets created in Dvaravati period that has been found at Nadune district, Mahasarakham province, northeastern Thailand. Many molds which are all terracotta have been used for production. Microstructure and composition of selected samples of the ancient amulets and the imitative one have been characterized using scanning electron microscopy combined with energy dispersive X-ray spectroscopy (SEM-EDS) and proton-induced X-ray emission spectroscopy (PIXE). The results show that many elements are present as major (C, O, and Si), minor (Na, Al, K, Ca, Fe, Cu, and Zn), and trace elements (Mg, P, S, Ti, Cr, and Mn). Quartz, rice husk, orthoclase, anhydrite, and laterite are the base components, while zinc, copper, and bauxite are the characteristic components. These non-destructive techniques show obvious composition difference between the ancient (original) and new (imitative) amulets.

1 Introduction

Most of people in Asia especially Thailand, Malaysia, Singapore, and China are Buddhists who believe in the sanctity of amulets. The sanctity includes a mystical power of providing happiness and fulfilling salvation of dangers. In Thailand, many ancient amulets are highly revered and demanded by people resulting in possession of very high prices and the production of imitations that are very similar to the originals.

Phra Kru Nadune is one of the ancient amulets created in Dvaravati period (7th - 11th century A.D.). It has been found at Nadune district, Mahasarakham province, northeastern

* Corresponding author: chatdanai.b@cmu.ac.th

Thailand. The amulet has been produced in many shapes including rectangle, curved top square, isosceles triangle, leaf triangle, square, and round-relief. Most of them are terracotta possessing different colors of texture including brown, yellow, red, gray, and pink. This possession is due to raw materials of laterite, rice husk, sand, and clay are burnt in an open-air furnace in which the temperature is unstable. Our previous works have been focused on the composition and morphology analyses of other Thai amulets such as Phra Somdej Wat Rakhang [1-2] and Phra Luang Pho Thuad [3]. In this work, the morphology and composition of Phra Kru Nadune amulets have been analyzed using scanning electron microscopy equiped with energy-dispersive X-ray spectroscopy (SEM-EDS) and protoninduced X-ray emission spectroscopy (PIXE) for examination of the imitative amulet.

2 Experimental

2.1 Sample preparation

8 samples of Phra Kru Nadune amulets (PND2, PND3, PND4, PND5, PND6, PND7, PND8, and PND9) with approximately 1-2 cm in base width and 5 cm in height were examined. These round-relief molded figures were produced in different batches with the same shape called "Nakprok Diao" which means "single Buddha image overspread with a great serpent" as shown in Figure 1.

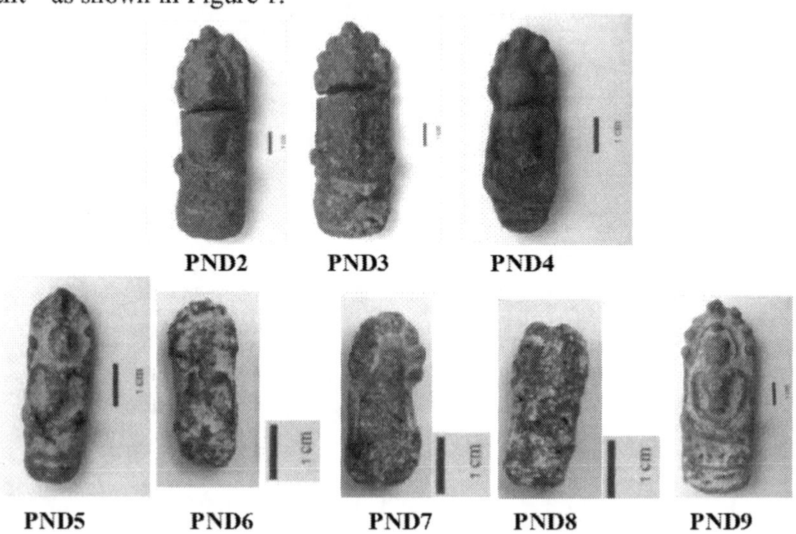

Fig. 1. Samples of Phra Kru Nadune amulets.

2.2 Characterization

2.2.1 SEM-EDS

Morphology and chemical composition were examined using Quanta 450 SEM equiped with X-Max EDS incorporated with INCA software. The operation was done with accelerating voltage of 5 kV. The chemical composition was determined in wt% and normalized to 100%. Each sample was probed in 3 positions of great serpent, chest, and base. The content of some trace element with approximately 0.01 wt% or lower was not able to be confirmed by SEM-EDS. This element was analyzed only by PIXE.

2.2.2 PIXE

PIXE was operated with 2-MeV-proton beam produced by 1.7 MV tandem Tandetron accelerator. The proton beam was collimated by a diaphragm of 1 mm in diameter in order to generate the beam current of 10 nA on sample. The characteristic X-ray was detected by Si(Li) detector. The sample was probed in the same positions (of great serpent, chest, and base) as examined by SEM-EDS. The chemical composition (for element with $Z \geq 11$) analyzed from PIXE spectra was performed using GUPIXWIN code [4-6]. The normalization of composition was done with 100%-x of the total amount of oxides, where the value of x was the sum of Na_2O and MgO determined by EDS. The subtraction of x was due to low accuracy in detection of Na and Mg by PIXE when compared to other elements with $Z \geq 13$.

The calibration of chemical compositions measured using PIXE and EDS was done before analysis of samples. The optical glasses with known chemical compositions were used as the references. It was found that the relative errors in calibration of the two techniques were less than 10%. The PIXE results can be therefore compared to EDS and the average content of each element from two techniques was evaluated [6]. Silicon with known composition was used as a reference for PIXE analysis as conducted in previous research [1-2,7-18]. The setup of samples for SEM-EDS and PIXE analyses are shown in Figure 2.

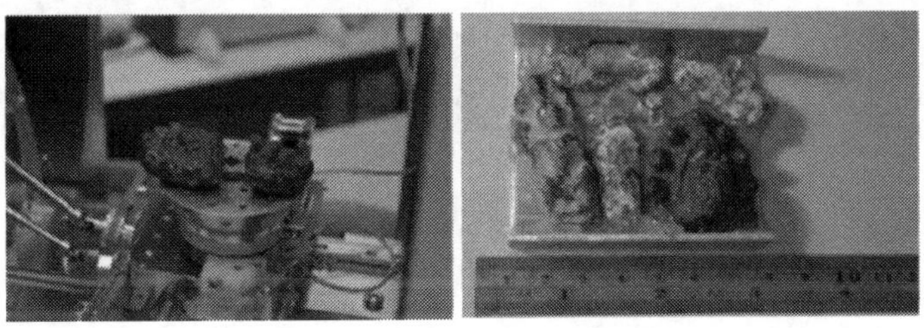

(a) **(b)**

Fig. 2. Setup of samples for (a) SEM-EDS and (b) PIXE analyses.

3 Results and Discussion

Morphologies of Phra Kru Nadune amulets are revealed by SEM images as shown in Figure 3, while the elemental compositions analyzed using EDS and PIXE are shown in Table 1. SEM micrographs show the surfaces with porous structures of cavities, cracks, irregular plate-like granules, and fine particles with a narrow range of size distribution. They also exhibit highly corroded surfaces resulted from environmental and chemical effects during their burial period in soil.

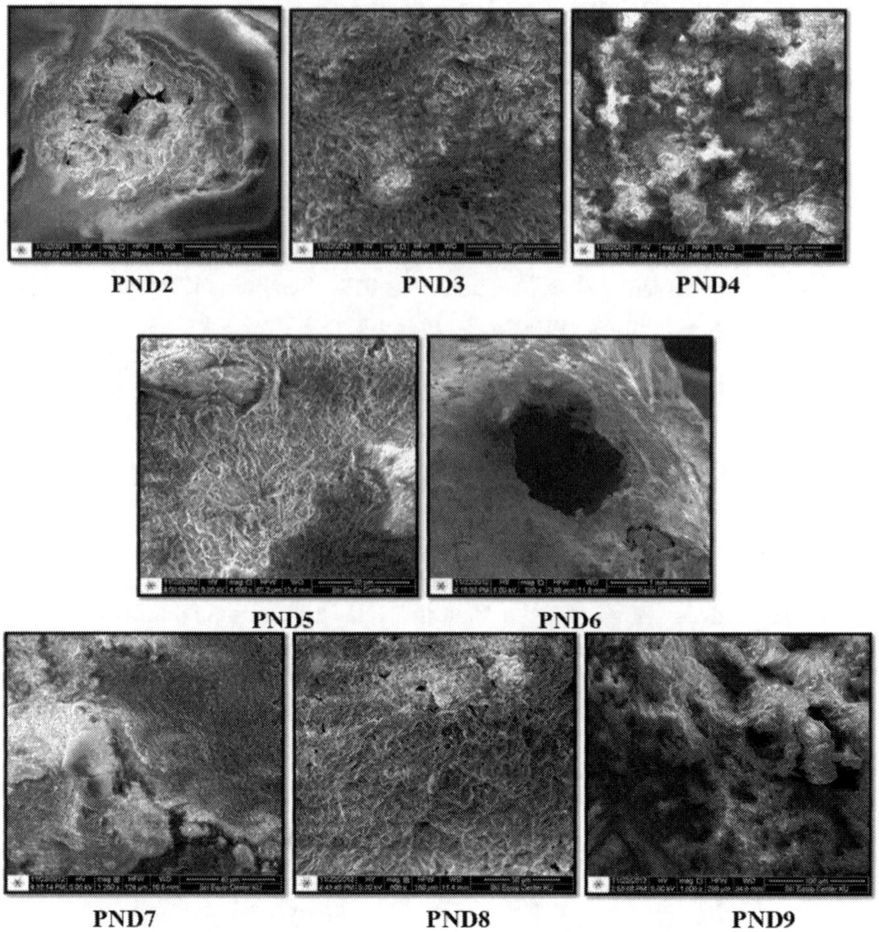

Fig. 3. Morphologies of samples of Phra Kru Nadune amulets.

The composition shown in Table 1 can be used to distinguish the amulets created from different batches of the same mold. The results show that C, O, and Si were the base composition of every samples. Some elements including Na, Mg, Si, K, and Ca were oxidized. SiO_2, Al_2O_3, CaO, S, and Fe were the components of minerals such as quartz, rice husk, orthoclase, anhydrite, laterite, and bauxite. The detection of P was proposed to be come from the burial environment, and minerals and bone ashes which were added to the clay paste [19]. The other oxides and elements (especially C) were present as impurities in raw materials of clay minerals and herb organics. However, B was only found with a high content (approximately 21 wt%) in sample PND3. It is proposed to be in the form of borax (used as a glaze in ceramic production in ancient time), boron silicate mineral (tourmaline), and/or tektite. Furthermore, the composition of sample PND9 was clearly different from the others by possession of larger amounts of Al_2O_3, Mn, and Fe, and smaller amounts of P, S, Cl, Cu, and Zn. The results show that PND9 was the imitative amulet.

It should be noted that the contents of elements such as Na, Mg, Al, Si, P, S, Cl, K, Ca, Ti, Fe, Cu, and Zn are the average of values derived from PIXE and SEM-EDS, trace elements such as Cr and Mn were determined only by PIXE, and the elements such as B, C, and O were only detected by SEM-EDS. The contents of some trace elements in some samples including S (PND9), Cl (PND9), K (PND6 and PND8), Cu (PND9), and Zn (PND9) were analyzed only by PIXE due to the elements were not detected by EDS. This indicates that the contents were beyond limit of detection of EDS. The high deviations of contents of carbon in every sample and those of SiO_2 and others in some samples reflect the high heterogeneity of materials texture of amulets. The high heterogeneity is proposed to be resulted from ancient indigenous technique of fabrication (~1,000 years ago) in which poor mixing of raw materials was done before shaping of amulet.

4 Conclusion

Non-destructive techniques of SEM-EDS and PIXE were essential for the analysis of Phra Kru Nadune amulets which provided information about the morphology and elemental composition. The results showed high heterogeneity of the amulet textures and composition difference between the ancient (original) and new (imitative) amulets. The results showed that the sample PND9 was the imitative amulet.

Table 1. Composition of Phra Kru Nadune amulets analyzed using SEM-EDS and PIXE.

Element	**Sample/Composition (wt%)**							
	PND2	**PND3**	**PND4**	**PND5**	**PND6**	**PND7**	**PND8**	**PND9**
B: Average	N/A *	21.67	N/A *	N/A *	N/A *	N/A *	N/A *	N/A *
SD		±3.42						
C: Average	27.47	57.55	60.04	42.75	39.41	74.57	62.46	34.57
SD	±5.72	±28.11	±31.66	±23.12	±12.56	±35.12	±25.23	±15.34
Na_2O: Average	1.97	1.78	0.50	0.63	2.68	0.59	3.17	1.11
SD	±0.74	±1.18	±0.19	±0.13	±1.57	±0.31	±1.62	±0.70
MgO: Average	0.77	0.25	0.22	2.71	1.70	0.64	0.42	1.71
SD	±0.36	±0.05	±0.09	±0.81	±1.01	±0.09	±0.29	±1.07
Al_2O_3: Average	0.51	1.78	0.32	5.50	0.79	4.31	1.02	11.77
SD	±0.41	±0.76	±0.19	±2.89	±0.24	±0.07	±0.78	±1.14
SiO2: Average	61.31	6.31	6.73	37.27	48.24	12.02	9.67	42.83
SD	±8.32	±3.40	±1.87	±6.37	±7.31	±4.57	±6.17	±16.85
P: Average	1.12	0.56	0.32	1.73	1.72	0.87	0.59	0.16
SD	±0.28	±0.41	±0.10	±1.10	±0.10	±0.40	±0.26	±0.08
S: Average	0.35	0.49	0.57	1.12	1.38	0.73	0.76	0.0817**
SD	±0.28	±0.29	±0.14	±0.35	±0.43	±0.40	±0.60	±0.0436
Cl: Average	0.44	0.32	0.38	0.26	0.28	0.69	0.13	0.0683**
SD	±0.29	±0.12	±0.19	±0.16	±0.07	±0.37	±0.07	±0.0487
K_2O: Average	0.19	0.24	0.20	0.17	0.0792**	0.25	0.0937	0.31
SD	±0.12	±0.05	±0.08	±0.05	±0.0513	±0.12	±0.0445	±.021
CaO: Average	0.52	2.14	1.12	2.56	1.61	3.39	3.43	1.22
SD	±0.47	±0.93	±1.06	±1.63	±0.91	±1.32	±1.74	±0.75
Ti: Average	0.20	0.18	0.15	0.16	0.16	0.22	0.24	0.16
SD	±0.18	±0.12	±0.09	±0.07	±0.11	±0.13	±0.11	±0.10
Cr: Average	0.0119**	0.0122**	0.0137**	0.0473**	0.0147**	0.0154**	0.0394*	0.0157**
SD	±0.0067	±0.0042	±0.0089	±0.0320	±0.0063	±0.0015	±0.0201	±0.0106
Mn: Average	0.0081	0.0078**	0.0163**	0.0227**	0.0261**	0.0046**	0.0269**	0.0943**
SD	±0.0019	±0.0024	±0.0086	±0.0046	±0.0128	±0.0013	±0.0115	±0.0527
Fe: Average	1.84	3.41	2.50	1.63	0.71	0.96	3.35	5.86
SD	±1.67	±0.81	±1.39	±0.92	±0.46	±0.67	±2.23	±2.97
Cu: Average	2.75	2.56	23.14	2.77	0.88	0.36	13.02	0.0226**
SD	±2.37	±0.61	±11.24	±1.43	±0.76	±0.22	±1.02	±0.0090
Zn: Average	0.54	0.74	3.78	0.67	0.32	0.38	1.58	0.0174**
SD	±0.45	±0.57	±2.59	±0.58	±0.21	±0.16	±0.63	±0.0064

N/A* denotes the content of element is below the detection limit of EDS (<0.01 wt%) and can not be analyzed using PIXE.

** denotes the element is analyzed only by PIXE.

This work is partially supported by Kasetsart University (Department of Earth Sciences at the Faculty of Science) and Chiang Mai University (Center of Excellence in Materials Science and Technology). We thank Science Equipment Center at Faculty of Science (Kasetsart University), and Plasma and Beam Physics Research Facility (Chiang Mai University) for SEM-EDS and PIXE measurements, respectively. Miss Woranan Hongrattanakorn was also thanked for analysis.

References

1. K. Won-in, S. Suksawang, S. Intarasiri, C. Thongleurm, T. Kamwanna, P. Dararutana, Advanced Materials Research **324**, 505 (2011).
2. K. Won-in, S. Suksawang, S. Intarasiri, C. Thongleurm, U. Tippawan, T. Kamwanna, P. Dararutana, J Radioanal Nucl Chem **294**, 38 (2012).
3. K. Won-in, C. Boonruang, K. Tepchai, P. Dararutana, *Microanalysis of Thai amulet: Phra Luang Pho Thuad*, in Book of Abstracts Siam Physics Congress 2021, 24-25 May 2021, Songkhla, Thailand (2021).
4. J.A. Maxwell, W.T. Teesdale, J.L. Campbell, Nucl. Instr. and Meth. B **95**, 407 (1995).
5. J.L. Campbell, N. Boyd, N. Grassi, P. Bonnick, J.A. Maxwell, Nucl. Instr. and Meth. B **268**, 3356 (2010).
6. N. Carmona, I. Ortega-Feliu, B. Gomez-Tubio, M.A. Villegas, Mater. Charact. **61**, 257 (2010).
7. K. Won-in, S. Pongkrapan, P. Dararutana, Advanced Materials Research **324**, 501 (2011).
8. K. Won-in, Y. Thongkam, S. Pongkrapan, S. Intarasiri, C. Thongleurm, T. Kamwanna, T. Leelawathanasuk, P. Dararutana, Spectrachimica Acta A **83**, 231 (2011).
9. K. Won-in, Y. Thongkam, T. Kamwanna, P. Dararutana J. Radioanal. Nucl. Chem. **294**, 247 (2012).
10. W. Dhanmanonda, K. Won-in, S. Tancharakorn, W. Tantanuch, C. Thongleurm, T. Kamwanna, P. Dararutana, IOP Conference Series: Materials Science and Engineering **37**, 012014 (2012).
11. K. Won-in, Y. Thongkam, S. Intarasiri, T. Kamwanna, P. Dararutana, IOP Conference Series: Materials Science and Engineering **37**, 012015 (2012).
12. K. Won-in, Y. Thongkam, W. Dhanmanonda, J. Dutchaneephet, T. Kamwanna, S. Intarasiri, S. Tancharakorn, W. Tantanuch, P. Dararutana, Physics Procedia **48**, 23 (2013).
13. K. Won-in, Y. Thongkam, W. Dhanmanonda, J. Dutchaneephet, T. Kamwanna, S. Intarasiri, S. Tancharakorn, W. Tantanuch, P. Dararutana, Physics Procedia **48**, 30 (2013).
14. P. Dararutana, K. Won-in, S. Intarasiri, T. Kamwanna, S. Tancharakorn, N. Sirikulrat, C.A. Hausenberger, Advanced Materials research **620**, 330 (2013).
15. K. Won-in, T. Sako, C. Thongleurm, S. Intarasiri, U. Tippawan, T. Kamwanna, W. Pattanasiriwisana, S. Tancharakorn, N. Kamonsuthipaijit, P. Dararutana, J. Radioanal. Nucl. Chem. **297**, 285 (2013).
16. K. Won-in, R. Supruangnet, P. Dararutana, Proc. of SPIE **11331**, 113310F1 (2020).
17. K. Won-in, P. Dararutana, Journal of Physics: Conference Series **1719**, 012075 (2021).

18. C. Boonruang, K. Won-in, P. Dararutana, Journal of Physics: Conference Series **1963**, 012038 (2021).
19. N. Nimsuwan, J. Dutchaneephet, P. Pakananan, P. Dararutana, K. Won-in, Malaysian Journal of Microscopy **15**, 83 (2019).

Simultaneous neutron and gamma imaging system for real time range and dose monitoring in Hadron Therapy and nuclear security applications

J. Lerendegui-Marco[1*] , *J.* Balibrea-Correa[1], *V.* Babiano-Suárez[1], *L.* Caballero[1], *C.* Domingo-Pardo[1], and *I.* Ladarescu[1]

[1]Instituto de Física Corpuscular, CSIC-University of Valencia, Spain

Abstract. GN-Vision is a novel dual γ-ray and neutron imaging system, which aims at imaging, simultaneously to the prompt gammas, the spatial origin of the slow and thermal neutron dose (<100 eV) generated during hadron therapy treatments. The proposed device can also be of interest for industrial applications as well as in nuclear security. The GN-Vision system has been designed following the technical developments of the iTED detector, an array of high efficiency Compton cameras intended for neutron-capture experiments, in which γ-ray energies span up to 5-6 MeV, similar to the energies encountered in hadron therapy. This manuscript presents the evolution of the i-TED detector towards the GN-Vision system and the first conceptual study of the simultaneous neutron and γ-ray imaging capability. Last, it reviews the status of the development and first results of the promising performance of this device for PG imaging in proton therapy, based on MC simulations.

1 Introduction

Hadron therapy presents clear advantages to conventional radiation therapy thanks to the maximum dose deposition at the end of the ion trajectory and its finite penetration in matter. However, it faces two important limitations related to neutron and gamma dose monitoring and ion-beam range verification, which hinder the potential benefits of this method. To solve these challenges, on-line dose and range diagnostic techniques are required.

For ion-range verification, only slit-cameras [1] have been applied for real-time Prompt γ-ray imaging (PGI) in clinical conditions [2]. Recently, the imaging of the secondary neutrons produced by the hadron-beam has been also proposed [3]. This technique is sensitive only to high energy neutrons (>20MeV), which are mainly emitted in the forward direction [4]. On the contrary, a dual neutron-gamma imaging device, such as the one proposed in this work, should be placed transversal to the beam in order to exploit Prompt γ-ray imaging.

* Corresponding author: jorge.lerendegui@ific.uv.es

Several groups have also explored PGI [5] and neutron detection [6] aiming at on-line dose monitoring. However, the latter provide no information about the spatial origin of neutrons. Dual neutrongamma prototypes represent a promising approach but most of the developed devices to date [7] are only sensitive to fast neutrons (>0.5 MeV) and, thus, a thermal and epithermal neutron monitor would provide complementary information. In addition, the size of such systems can also be a limitation for clinical treatments.

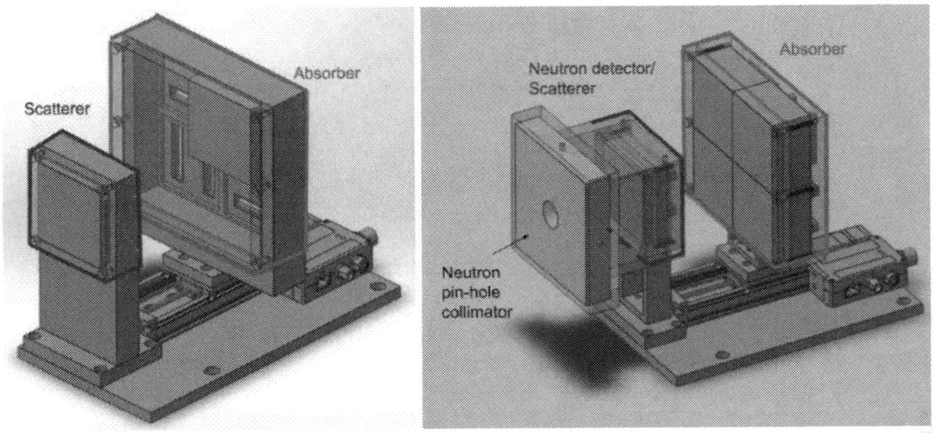

Fig. 1. Technical implementation of the i-TED Compton camera (left) and evolution towards the proposed GN-Vision device (right). The main components of both devices have been highlighted. The dimensions are indicated in Table 1.

Real-time imaging of γ-rays and neutrons is also of interest for remotely-controlled measurements in nuclear accidents [8], nuclear security applications and non-proliferation inspections [9]. In this field, the existing systems with dual γ-ray and neutron imaging capabilities consist of bulky scintillation detector arrays [10] able to image only fast neutrons. However, in many applications the radioactive material is covered by moderating material, thus leading to a thermal neutron spectrum.

The size of these devices also represents a clear disadvantage in terms of portability and applicability.

In this context, a new compact simultaneous gamma and neutron imaging device, hereafter called GN-Vision, is proposed to tackle some of the most relevant requirements in the aforementioned applications.

2 The GN-Vision concept and its evolution from i-TED

The system consists of a compact and portable device capable of measuring and simultaneously imaging both thermal and slow-neutrons and γ-rays with high efficiency. Simultaneous gamma and neutron imaging systems should fulfill several aspects [11]: being sensitive and able to discriminate the two particles, featuring position sensitivity to determine the interaction positions and exploit either electronic or passive collimation techniques to carry out imaging of the incoming particles or radiation.

The GN-Vision device has been developed as a a modification of the i-TED Compton camera developed for neutron capture time-of-flight experiments [12–14]. The technical implementation of i-TED, displayed in the left panel of Fig. 1 has been modified in such a way that the capability to image γ-rays is complemented in the same device with the imaging of neutrons. The first prototype of the proposed GN-Vision is displayed in the right panel of Fig. 1.

The proposed dual imaging device follows the next working principle:

- The Compton imaging technique [15] is applied to image γ-rays with energies between 100 keV and several MeV using two position-sensitive detection planes that are named as Scatterer and Absorber in Fig. 1.

Table 1. Summary of the main components and imaging techniques used in i-TED detector and its evolution GN-Vision.

	i-TED	GN-Vision
First plane	$LaCl_3$ 50×50 × 15 mm^3	CLYC-6 50×50 × 10 mm^3
Second plane	4 $LaCl_3$ crystals 50×50 × 25 mm^3	4 $LaCl_3$ crystals of 50×50 × 25 mm^3
γ-ray imaging	Compton (active)	Compton (active)
Neutron imaging	-	Pinhole camera (passive)

- In GN-Vision, unlike in the early i-TED detector, the first detection plane is capable of discriminating γ-rays and neutrons and fully absorbing the latter for energies below 1 keV.
- A passive neutron collimation system (see pin-hole collimator in Fig. 1) coupled to the neutron detecting plane enables the imaging of neutrons with energies below 100 eV using a similar concept as pin-hole γ-ray cameras [16].

The main features and components of GN-Vision are compared to the previous i-TED in Table 1. For the imaging of γ-rays, the GN-Vision operates, exactly as in the i-TED concept, as a Compton camera consisting of two position sensitive detection planes [12, 15]. The reader is referred to Refs. [13, 14, 17] for more details on the development, characterization and performance of the i-TED detector. In order to achieve the imaging of neutrons, the $LaCl_3$ crystal of the first detection layer of i-TED is replaced by a Cs_2LiYCl_6:Ce scintillation crystal enriched with ^{6}Li at 95%(CLYC-6), able of discriminating γ-rays, fast and thermal neutrons by Pulse Shape Discrimination (PSD) [18]. The use of a CLYC-6 scintillator as Scatterer detector ensures also sufficient energy resolution for Compton imaging in GN-Vision. As shown in Fig. 1, a neutron absorbing pin-hole collimator made of ^{6}Li-enriched Li-Polyethylene (6LiPE) is attached to the first plane to enable the reconstruction of 2D neutron-images. On the other hand, the 6LiPE material does not interfere with γ-rays above ~100keV, thereby enabling the simultaneous neutron-gamma vision within the same device. The first design study of this collimator by means of MC simulations is discussed in Sec. 3. More details on the technical implementation of GN-Vision are given in Ref. [19].

3 Monte Carlo conceptual design of GN-Vision

The first conceptual design of the GN-Vision system has been carried out by means of MC simulations using the GEANT4 toolkit (v10.6) [20] including the G4NeutronHP package [21] for the accurate simulation of neutron interactions. This study aimed at demonstrating and optimizing the neutron vision concept while retaining the high γ-ray imaging efficiency and resolution of the Compton camera. The geometry model implemented in Geant4 is shown in Fig. 2. In this figure, we indicate the critical parameters in the design of the neutron collimator: diameter (D), thickness (T) and focal distance (F).

 https://doi.org/10.1051/epjconf/202226105001

The first study of the neutron imaging capability consisted on simulating three isotropic point-like neutron sources of energies ranging from thermal energies (25 meV) to 1 keV located at 20 cm from the collimator and separated from each other by 5 cm (see Fig. 2). The simple geometry of Fig. 2 served to optimize the aforementioned design parameters of the pin-hole collimator for various neutron energies below 1 keV. A more detailed description of these simulations is given in Ref. [19].

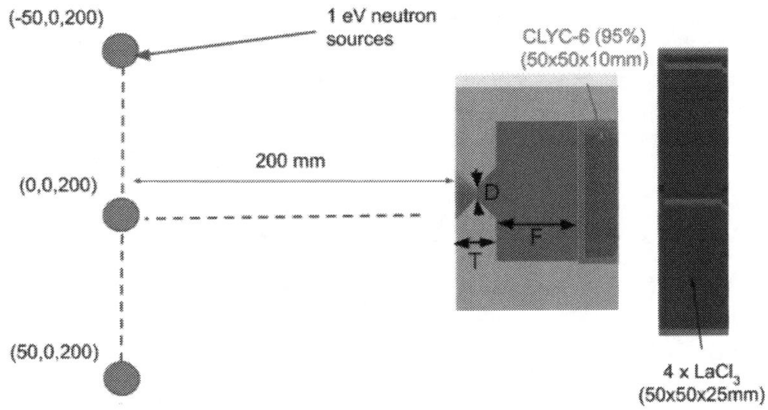

Fig. 2. Sketch of the geometry model of GN-Vision implemented in Geant4 for the conceptual demonstration of the neutron imaging capability.

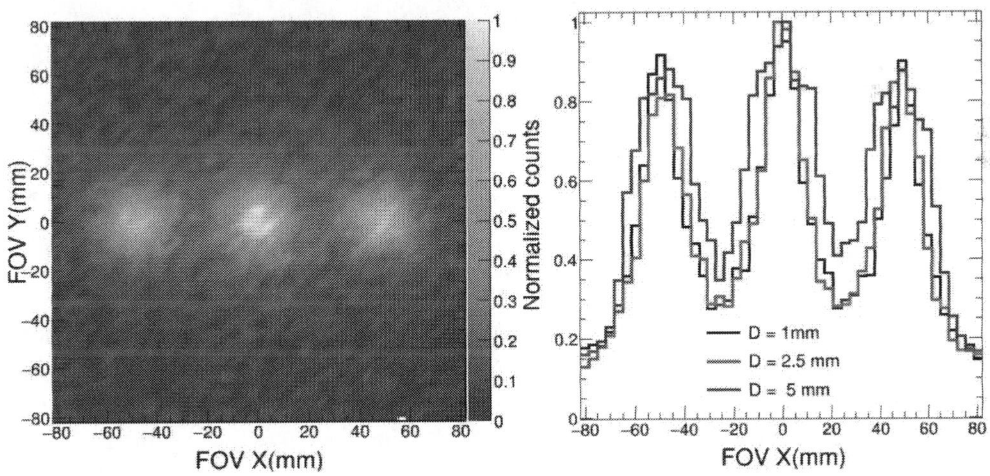

Fig. 3. Reconstructed neutron images (left) for three isotropic point-like sources of 1 eV neutrons shown in the sketch of Fig. 2 and projections (right) along the X-axis of the images obtained with different values of the collimator aperture (D) for a fixed thickness of T = 20 mm and a focal distance F = 40 mm.

The left panel of Fig. 3 shows the first results of the neutron images obtained from the simulations of the GN-Vision concept corresponding to an neutron energy of 1 eV. The balance between efficiency and resolution in the neutron images is extremely dependent on the aperture (D) of the pin-hole collimator, as it is illustrated in the 1D projections of the images shown in the right panel of Fig. 3. The full study of the neutron collimator design and its impact in the γ-ray Compton images can be found in Ref. [19].

4 Perspectives for Prompt Gamma Imaging

Moving a step forward towards the medical applications of the GN-Vision, the prospects of the predecessor i-TED detector for range verification in proton therapy via Prompt Gamma (PG) imaging has been recently studied by means of MC calculations [22]. In this work, a pencil-beam of 120 MeV protons was impinging on a water phantom with a size of $10\times10\times20$ cm.

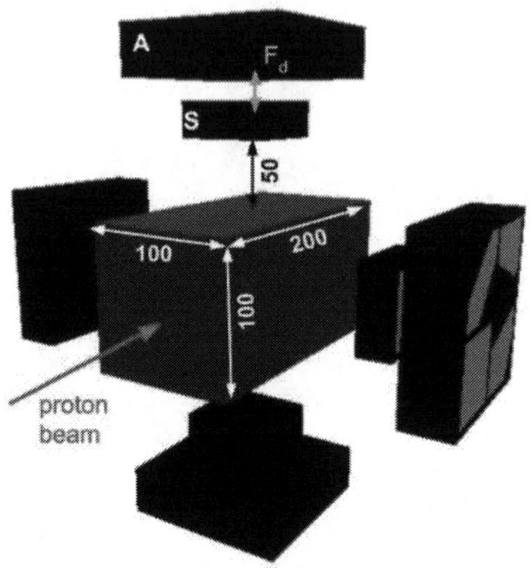

Fig. 4. Simple geometry model implemented in the GEANT4 code to study the potential application of i-TED/GN-Vision to Prompt Gamma imaging in proton therapy.

Figure 4 shows a schematic view of the simulated geometry. The phantom is surrounded by four i-TED Compton cameras at 50 mm distance from each lateral surface and at a depth of 100 mm. The Prompt gamma radiation emitted along the track of the protons, closely correlated with the proton range, was imaged using the Compton technique with these devices. All the details on these simulations can be found in Ref. [22].

To illustrate the promising results of this MC study, we show in Figure 5 the reconstructed depth profile obtained for the 4.4 MeV prompt γ-rays originated by the protons along the water phantom. The right panel of the figure shows reconstructed 2D Compton image, where one clearly appreciates the spatial distribution of the emitted PG.

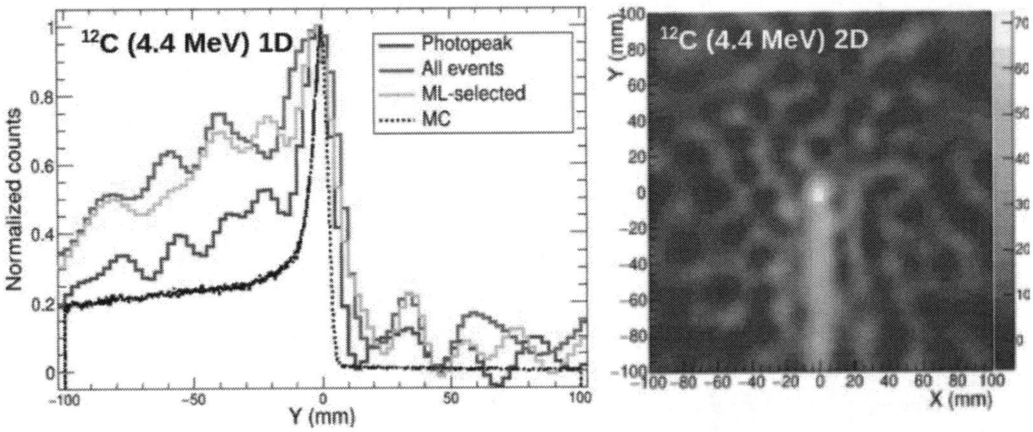

Fig. 5. Reconstructed Compton image for the 4.4 MeV prompt γ-rays (right) and 1D distribution along the proton beam axis comparing the true depth distribution (MC) with the profiles reconstructed from the Compton images (left). See text for details.

An accuracy of few mm is achieved in the reproduction of the PG emission fall-off as shown in the left panel of Fig. 5, where the reconstructed profiles are compared with the MC true distribution (black). Our inital reconstruction of the depth profile (red) has been improved thanks to a Machine Learning (ML) identification of full-energy events developed in this work [22]. As a consequence, one can recover partially the resolution and signal-to-background ratio of an ideal Camera with only full-energy events, shown by the blue line.

5 Summary and outlook

In this work we have presented the first conceptual design of the GN-Vision system, a compact simultaneous neutron-gamma imager, which may have potential applications for range and dose monitoring in HT and in nuclear security.

The GN-Vision device, designed as an evolution of the early i-TED detector, consists of two position sensitive detection layers based on CLYC-6 and $LaCl_3$ detector to enable Compton γ-ray imaging. A neutron pinhole collimator attached to the first enables the imaging of slow neutrons.

The first results of the MC design study of the system, focused in the optimization of the collimator parameters, demonstrate the capability of imaging neutrons with energies below 100 eV.

The imaging capability of GN-Vision is experimentally demonstrated with the predecessor iTED and the first MC study of its potential applicability for range verification in proton therapy shows promising perspectives. As for the neutron vision, after the conceptual proof-of-concept, the first experimental tests will be carried out in the upcoming months. The system has been recently patented [23], and further R&D will follow in the next years.

This work has been carried out in the framework of a project funded by the European Research Council (ERC) under the European Union's Horizon 2020 research and innovation programme (ERC Consolidator Grant project HYMNS, with grant agreement No. 681740). The authors acknowledge support from the Spanish Ministerio de Ciencia e Innovación under grants PID2019-104714GB-C21, FPA2017-83946-C2-1-P, FIS2015-71688-ERC, CSIC for funding PIE-201750I26.

References

1. J. Smeets, F. Roellinghoff, D. Prieels, F. Stichelbaut, A. Benilov, P. Busca, C. Fiorini, R. Peloso, M. Basilavecchia, T. Frizzi et al., Physics in Medicine and Biology **57**, 3371 (2012)
2. C. Richter, G. Pausch, S. Barczyk, M. Priegnitz, I. Keitz, J. Thiele, J. Smeets, F.V. Stappen, L. Bombelli, C. Fiorini et al., Radiotherapy and Oncology **118**, 232 (2016)
3. K.S. Ytre-Hauge, K. Skjerdal, J. Mattingly, I. Meric, Scientific Reports **9**, 2011 (2019)
4. H. Iwase, et al., Radiation Protection Dosimetry **126**, 615 (2007)
5. G. Llosá, M. Trovato, J. Barrio, A. Etxebeste, E. Muñoz, C. Lacasta, J.F. Oliver, M. Rafecas, C. Solaz, P. Solevi, Frontiers in Oncology **6**, 14 (2016)
6. J. Farah, V. Mares, M. Romero-Expósito, S. Trinkl, C. Domingo, V. Dufek, M. Klodowska, J. Kubancak, Kneževic, M. Liszka et al., Medical Physics′ **42(5)**, 2572 (2015)
7. S.D. Clarke, E. Pryser, B.M. Wieger, S.A. Pozzi, R.A. Haelg, V.A. Bashkirov, R.W. Schulte, Medical Physics **43**, 5915 (2016)
8. K. Vetter, R. Barnowksi, A. Haefner, T.H. Joshi, R. Pavlovsky, B.J. Quiter, Nuclear Instruments and Methods in Physics Research Section A: Accelerators, Spectrometers, Detectors and Associated Equipment **878**, 159 (2018)
9. A. Poitrasson-Rivière, J.K. Polack, M.C. Hamel, D.D. Klemm, K. Ito, A.T. McSpaden, M. Flaska, S.D. Clarke, S.A. Pozzi, A. Tomanin et al., Nuclear Instruments and Methods in Physics Research Section A: Accelerators, Spectrometers, Detectors and Associated Equipment **797**, 278 (2015)
10. S.P. et al., *Dual particle imaging system for standoff snm detection in high-background radiation environment* (US2012/0256094A1 Oct. 2012, Priority: 6.4.2012)
11. H. Al Hamrashdi, S.D. Monk, D. Cheneler, Sensors **19** (2019)
12. C. Domingo-Pardo, Nuclear Instruments and Methods in Physics Research A **825**, 78 (2016)
13. V. Babiano, J. Balibrea, L. Caballero, D. Calvo, I. Ladarescu, J. Lerendegui, S. Mira Prats, C. Domingo-Pardo, Nuclear Instruments and Methods in Physics Research A **953**, 163228 (2020), 1908.08533
14. V. Babiano, J. Lerendegui-Marco, et al., The European Physical Journal A **57**, 197 (2021)
15. D. Everett, Proceedings of the Institution of Electrical Engineers **124**, 995 (1977)
16. H.O. Anger, Review of Scientific Instruments **29** (1958)
17. J. Balibrea-Correa, J. Lerendegui-Marco, V. Babiano-Suárez, L. Caballero, D. Calvo, I. Ladarescu, P. Olleros-Rodríguez, C. Domingo-Pardo, Nuclear Instruments and Methods in Physics Research Section A: Accelerators, Spectrometers, Detectors and Associated Equipment **1001**, 165249 (2021)
18. A. Giaz, L. Pellegri, F. Camera, N. Blasi, S. Brambilla, S. Ceruti, B. Million, S. Riboldi, C. Cazzaniga, G. Gorini et al., Nuclear Instruments and Methods in Physics Research Section A: Accelerators, Spectrometers, Detectors and Associated Equipment **810**, 132 (2016)

19. J. Lerendegui-Marco, J. Balibrea-Correa, V. Babiano-Suárez, L. Caballero, D. Calvo, I. Ladarescu, C. Domingo-Pardo, Nuclear Instruments and Methods in Physics Research A (**submitted**) (2021)
20. J. Allison, K. Amako, J. Apostolakis, P. Arce, M. Asai, T. Aso, E. Bagli, A. Bagulya, S. Banerjee, G. Barrand et al., Nuclear Instruments and Methods in Physics Research Section A: Accelerators, Spectrometers, Detectors and Associated Equipment **835**, 186 (2016)
21. E. Mendoza, D. Cano-Ott, T. Koi, C. Guerrero, IEEE Transactions on Nuclear Science **61**, 2357 (2014)
22. J. Lerendegui-Marco, J. Balibrea-Correa, V. Babiano-Suárez, I. Ladarescu, C. Domingo-Pardo, Scientific Reports **(submitted)** (2021)
23. *Device for simultaneous detector, identification, quantification and/or location of gamma radiation and neutron source* (International Patent PCT/ES2021/070342)

First in-beam tests on simultaneous PET and Compton imaging aimed at quasi-real-time range verification in hadron therapy

Javier Balibrea-Correa[1*], *Jorge* Lerendegui-Marco[1], *Victor* Babiano-Suarez[1], *César* Domingo-Pardo[1], *Ion* Ladarescu[1], *Carlos* Guerrero[2,3], *Teresa* Rodríguez-González[2,3], *Maria del Carmen* Jiménez-Ramos[2,3], and *Begoña* Fernández-Martínez[2,3]

[1]Instituto de Física Corpuscular, CSIC-University of Valencia, Valencia, Spain
[2]University of Seville, Seville, 41012 Spain
[3]Centro Nacional de Aceleradores (U. Sevilla, CSIC, Junta de Andalucía), Sevilla, 41092 Spain.

Abstract. Hadron therapy with protons has advantages with respect to conventional radiotherapy because of the maximization of the dose at the Bragg peak. As a drawback, and because of different systematic uncertainty sources, a quasi-real time monitoring for the proton range verification is required to reduce safety margins. In this respect, two gamma-ray imaging techniques are pursued: prompt gamma-ray monitoring and positron-annihilation tomography (PET). The promising prompt gamma-ray monitoring requires detection systems with large detection efficiency, high time resolution, compactness, fast response, low sensitivity to neutron-induced backgrounds and powerful image reconstruction capabilities. On the other hand, in-beam PET surveys require additionally good γ-ray position reconstruction resolution. In this contribution we show that, to a large extent, both approaches can be simultaneously accomplished by using an array of Compton cameras conveniently arranged around the target volume. Here we demonstrate experimentally the suitability of such an array, named i-TED, for PG monitoring in ion-range monitoring during Hadron Therapy, in-beam PET survey and β^+ production yield measurements capability. Furthermore, with the use of GPUs, a quasi-real time PG monitoring and in-beam PET can be achieved.

1 Introduction

Over the last decades Hadron Therapy (HT) has gained popularity for treatment of certain type of cancers. This increasing trend is reflected on the more than 220000 patients treated using protons by 2019 worldwide, the 111 HT centers currently in operation and the more than 50 under construction or at planning stages [1, 2].

The benefits from HT are larger compared to the conventional radiotherapy, targeting precisely the tumor area and minimizing the damage into neighbouring tissues.

*Corresponding author: javier.balibrea@ific.uv.es

As a consequence, these treatments decrease the probability of secondary cancers and other related issues derived from conventional radiotherapy [3]. However, as a drawback, different sources of systematic uncertainties arise in HT affecting to the treatment plans, thus increasing the safety margins and reducing the potential aforementioned benefits. The most critical uncertainties sources are anatomical changes, patient setup errors and uncertainty of proton stopping powers in different materials [4].

Ion range verification, and if possible on-line, is one of the keys to reduce those safety margins using imaging techniques and secondary particle produced in nuclear reactions during the proton irradiations. Two main imaging techniques can be used either alone or combined: Prompt Gamma-ray (PG) monitoring [5] and In-vivo PET survey [3]. PG monitoring takes advantage of the high spatial correlation from the quasi-instantaneous secondary γ-ray particles emitted just after the nuclear reactions occur during the irradiation, such as $^{16}O(p,x\gamma)^{12}C$, $^{12}C(p,p')^{12}C$,$^{16}O(p,x\gamma)^{15}N$. The monitoring is made by means of collimated γ-ray detection systems and/or Compton cameras during the patient irradiation. However, the harsh background conditions and large counting rates registered by the detection systems during the proton irradiations introduces large measurement complications. In the other hand, in-vivo PET monitoring benefits from the relative long-lived isotopes produced during the irradiation that decays by β^+ emission (^{11}C, ^{13}N,^{15}O...). After the proton irradiations, the patients are moved to a commercial PET machine. Compared to the PG monitoring, the detection systems suffers from less counting rate and background problems. However, the biological wash-out deteriorate the accuracy of the image, and therefore the precision of the dose assessment.

While the imaging is a well known methodology, the amount of secondary particles and therefore of the dose assessment strongly rely on the quality of nuclear data, particularly on the nuclear reactions cross-section data and β^+ yields in the proton energy range of interest for HT, usually tuned with Monte Carlo simulation codes to match the experimental data [6]. Thus, this is a strong indication that new nuclear data is needed to improve the dose and ion verification accuracy.

In this work we have tested the viability of a dual Compton imager and in-beam PET detection system for PG monitoring during the proton irradiations. This study was conducted in parallel to a survey for β^+ cross-section measurements as a function of the proton energy [7].

2 Experimental setup

The experimental apparatus used for this first in-beam test was a part of the so called iTED detection system [8], an advanced array of two-planes Compton cameras specifically designed for (n,γ) cross-section measurements using the Time-Of-Flight technique. In such experiments, the γ-ray yield from the radiative capture reactions of interest is rather low, when compared to the neutron-induced background levels in the experimental area. Additionally, the γ-ray energies from those reactions covers from a few keV up to 5-6 MeV, matching well with the γ-ray lines produced during HT treatments. Each i-TED module was therefore designed aiming at largest possible detection efficiency, while minimizing its sensitivity to neutron-induced γ-ray backgrounds. With that objective in mind we implemented large $LaCl_3(Ce)$ monolithic scintillation crystals optically coupled to position sensitive Silicon Photomultipliers. For further details of the i-TED modules the reader is referred to [8– 12].

Given the similarities in experimental conditions between neutron capture cross-section experiments and proton therapy treatments, it becomes of interest to investigate the suitability

 https://doi.org/10.1051/epjconf/202226105002

of the developed i-TED array also for ion-range verification studies. Thus, a first in-beam test experiment was carried out at the Cyclotron facility of the Centro Nacional de Aceleradores situated in Seville [13]. The accelerator delivers a 18 MeV proton beam to the dedicated radiobiological experimental bunker with currents up to ~80 μA [14, 15]. The experimental setup placed in the bunker was made of two i-TED modules (i-TED.A and i-TED.B), front

Figure. 1. Picture of the experimental setup. The different parts such as detectors, samples holder and others are labelled for a comprehensive understanding of the experimental arrangement.

to front and covering a 100×100 mm^2 PET Field of View (FoV), as it is displayed in the picture of Fig. 1. A sample holder was placed between both i-TED modules and fully aligned with entrance of the proton beam. Five thin samples or targets (41.2×41.2×0.8 mm^3) were situated in the slots of sample holder with a regular gap of 1.6 cm. At the end of the holder a thick graphite layer (2 mm thick) was placed in order to fully stop the beam and read the delivered proton current. A remotely movable PLA matrix, acting as β^+ converter, completed the experimental setup of this first test experiment. The sample holder together with the movable part and the thin samples makes a 5.2 × 5.5 × 10.3 cm^3 solid piece.

Three different configurations were used during the experiment aiming to cover different material yiels and proton energies. In the first configuration the thin targets consisted of Nylon material. In the second one, in addition to the Nylon foils, a proton energy degrader was placed upstream in order to decrease the incoming proton energy. Finally, in the last configuration, the Nylon targets was replaced by PMMA material.

The experiment was carried out in two separated stages consisting of beam-on and beam-off phases. During the first stage, the β^+ converter is removed from the proton beam path, thus irradiating with protons all the thin layers under study. During the beam-off phase the PLA converter was inserted into the samples holder, thereby filling up the gap between thin layers and reducing the mean free path of the e$^+$ particles from the nuclear decays.

3 Experimental procurement and results

The individual detectors were calibrated in energy using a ^{152}Eu standard calibration source together with the γ-ray lines observed during the proton irradiation. This approach ensured a reliable calibration for the individual detectors up to at least 4.5 MeV. Details about the γ-ray reconstruction and the implemented Machine Learning algorithms are explained in [11]. The experimental PG images were reconstructed for each i-TED Compton module individually (see Fig. 1.). To this aim an analytical inversion formula of the Compton law was used, following the prescription given in Ref. [16] and using the 4.4 MeV γ-ray line produced mainly in $^{12}C(p,p')$ reactions. For PET imaging an analytical algorithm was used, which was based on the geometrical intersection with the image plane of 511 keV γ-ray pairs emitted in

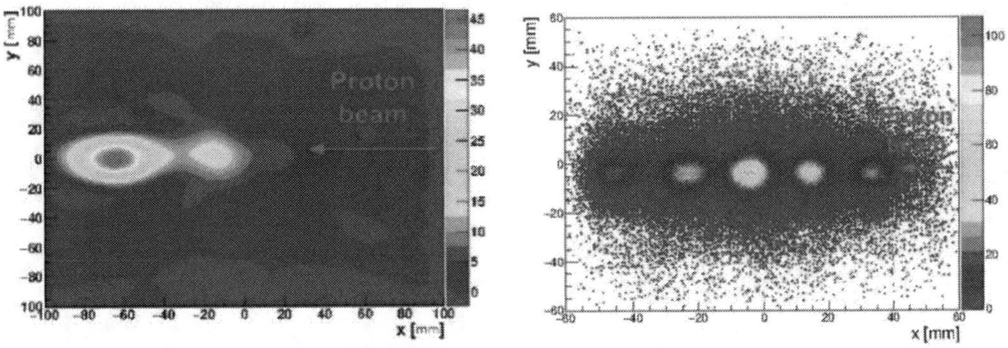

Figure. 2. Left panel: PG monitoring using Compton imaging for i-TED-B during the beam-on phase of the Nylon proton irradiation. Right Panel: PET imaging obtained during the Beam-off period of the same configuration.

back-to-back directions as reconstructed with the two i-TED modules operated face-to-face and in time-coincidence (see Fig. 1.).

It is worth mentioning that in order to accomplish a quasi-real-time monitoring, all the γ-ray position reconstruction in the individual detectors and the Compton imaging reconstruction algorithms were implemented in GPUs using CUDA [17], taking advantage of a speed-up factor of ~300 compared to the same single-thread reconstruction code version.

Owing to space limitations, only the experimental results for the first configuration are presented here. The left panel of Fig 2. shows the PG monitoring of i-TED.B during the beam-on phase. As expected, the reconstructed Compton image is dominated by the thick graphite layer due to the large amount of Carbon material placed there. Additionally, a second peak is observed in the center of the image, where the maximum Compton detection efficiency of the setup is located, corresponding with the position of the central Nylon layer.

The right panel of Fig. 2. displays the PET reconstructed image during the beam-off phase. The five Nylon foils are reconstructed with high spatial resolution, covering almost 100% of the PET field of view in the transversal direction.

4 Summary and outlook

In this first pilot experiment we have demonstrated the suitability of two high-efficiency Compton cameras (i-TED modules) for simultaneous proton-range verification and in-beam PET imaging. The data obtained from this work will be used in another study [7] to extract valuable β^{+} production cross-sections at low proton energies. The use of algorithms implemented in GPUs guarantee an almost quasi-real-time monitoring thanks to the ~300 speed-up factor. An article is in preparation [18], which describes in detail the experimental setup, methodology, data reduction and results for all the different configurations.

Future experiments using a similar experimental setup are planned at the Heidelberg Ion Beam Therapy Center (HIT) for covering the proton energy of interest in HT with protons.

This work has been carried out in the framework of a project funded by the European Research Council (ERC) under the European Union's Horizon 2020 research and innovation programme (ERC Consolidator Grant project HYMNS, with grant agreement No. 681740).

References

[1] H. Youngyih, Radiat Oncol J 37, 232 (2019), http://www.e-roj.org/journal/view.php?number=1424

[2] P.T.C.O. Group, *Particle therapy co-operative group*, https://www.ptcog.ch/ (2021)

[3] A.C. Knopf, A. Lomax, Physics in Medicine and Biology 58, 131 (2013), https://doi.org/10.1088/0031-9155/58/15/r131

[4] H. Paganetti, Physics in Medicine and Biology 57 (2012), https://doi.org/10.1088/0031-9155/57/11/R99

[5] J. Krimmer, D. Dauvergne, J. Létang, Testa, Nuclear Instruments and Methods in Physics Research Section A: Accelerators, Spectrometers, Detectors and Associated Equipment 878, 58 (2018)

[6] M. Durante, H. Paganetti, Reports on Progress in Physics 79, 096702 (2016)

[7] Rodríguez-González, T., Guerrero, C., Jiménez-Ramos, M.C., Dendooven, P., Lerendegui-Marco, J., Fraile, L.M., Millán-Callado, M.A., Ozoemelam, I., Parrado, A., Quesada, J.M., EPJ Web Conf. 239, 24003 (2020)

[8] C. Domingo-Pardo, Nuclear Instruments and Methods in Physics Research Section A: Accelerators, Spectrometers, Detectors and Associated Equipment 825, 78 (2016)

[9] P. Olleros, L. Caballero, C. Domingo-Pardo, V. Babiano, I. Ladarescu, D. Calvo, P. Gramage, E. Nacher, J. Tain, A. Tolosa, Journal of Instrumentation 13, P03014 (2018)

[10] V. Babiano, J. Balibrea, L. Caballero, D. Calvo, I. Ladarescu, J. Lerendegui, S. Mira Prats, C. Domingo-Pardo, Nuclear Instruments and Methods in Physics Research Section A: Accelerators, Spectrometers, Detectors and Associated Equipment 953, 163228 (2020)

[11] J. Balibrea-Correa, J. Lerendegui-Marco, V. Babiano-Suárez, L. Caballero, D. Calvo, I. Ladarescu, P. Olleros-Rodríguez, C. Domingo-Pardo, Nuclear Instruments and Methods in Physics Research Section A: Accelerators, Spectrometers, Detectors and Associated Equipment 1001, 165249 (2021)

[12] V. Babiano-Suárez, J. Lerendegui-Marco, J. Balibrea-Correa, L. Caballero, D. Calvo, I. Ladarescu, D. Real, C. Domingo-Pardo, F. Calviño, A. Casanovas et al., The European Physical Journal A 57, 197 (2021)

[13] C. Universidad de Sevilla, *Centro nacional de aceleradores*, http://cna.us.es/index.php/es/ (2020)

[14] A. Baratto-Roldán, M.d.C. Jiménez-Ramos, M.C. Battaglia, J. García-López, M.I. Gallardo, M.A. Cortés-Giraldo, J.M. Espino, Instruments 2 (2018)

[15] A. Baratto-Roldán, M. del Carmen Jiménez-Ramos, S. Jimeno, P. Huertas, J. García-López, M.I. Gallardo, M.A. Cortés-Giraldo, J.M. Espino, Physica Medica 74, 19 (2020)

[16] T. Tomitani, M. Hirasawa, Physics in Medicine and Biology 47, 2129 (2002)

[17] J. Nickolls, I. Buck, M. Garland, K. Skadron, Queue 6, 40–53 (2008)

[18] J. Balibrea-Correa, et al., In preparation (2022)

A new route to produce ^{52g}Mn with high purity for MultiModal Imaging

Mario Pietro Carante[1,2*], *Francesca* Barbaro[1,3], *Luciano* Canton[3], *Alessandro* Colombi[1,2], and *Andrea* Fontana[2]

[1]Università di Pavia, Dipartimento di Fisica, 27100 Pavia, Italy
[2]INFN (Istituto Nazionale di Fisica Nucleare), Sezione di Pavia, 27100 Pavia, Italy
[3]INFN (Istituto Nazionale di Fisica Nucleare), Sezione di Padova, 35131 Padova, Italy

Abstract. The ^{52g}Mn radionuclide is suitable for the innovative MultiModal Imaging technique, and in particular for a PET/MRI scan, due to its physical properties. The standard cyclotron-based production of ^{52g}Mn relies on the nuclear reaction ^{Nat}Cr(p,x)^{52g}Mn, but we have investigated theoretically the possibility of an alternative and competitive route, the reaction ^{Nat}V(α,x)^{52g}Mn, which has not been considered for this purpose so far. By using the nuclear reaction code TALYS, we found some discrepancies between the theoretical calculations of the cross sections and the corresponding experimental data. Therefore we tuned the parameters governing the nuclear level densities in the microscopic models implemented in TALYS, thus improving the agreement with the data. Then, by studying the cross sections for ^{52g}Mn and its contaminants, we have identified an optimal energy window for the production of high purity ^{52g}Mn, around 40 MeV. We have also calculated the time evolution of the number of nuclei of the different Mn isotopes, for an irradiation in this energy window, finding that this route is expected to lead to a higher yield and Radionuclidic Purity with respect to the standard reaction with ^{Nat}Cr. The study suggests the reaction ^{Nat}V(α,x)^{52g}Mn as a promising alternative route for the production of ^{52g}Mn.

1 Introduction

MultiModal Imaging is an innovative diagnostic technique consisting in the combination of two different diagnostic exams, based on different physical processes. An example is represented by the combination of Positron Emission Tomography (PET) and Magnetic Resonance Imaging (MRI), performed simultaneously by using a singular molecular agent. The radionuclide ^{52g}Mn appears of particular interest for this purpose because of its physical properties: it decays by the emission of a positron with energy of about 0.6 MeV, suitable to perform PET with good spatial resolution [1], and it has paramagnetic properties, which make it suitable for MRI. Moreover, its relatively long half-life of about 5.6 days is suitable for the radiolabeling of antibodies and other slow biological compounds for the study of pharmacokinetics.

* Corresponding author: mariopietro.carante@unipv.it

The standard cyclotron-based production route of ^{52g}Mn relies on the nuclear reaction ^{Nat}Cr(p,x)^{52g}Mn, which leads to the co-production of long-lived contaminants ^{53}Mn and ^{54}Mn: the latter in particular is the most important contaminant in view of clinical applications. Indeed, ^{48}Mn, ^{49}Mn, ^{50}Mn, ^{51}Mn and ^{52m}Mn have half-lives smaller than 1 h, and their contamination, if any, is thus negligible after a few hours. On the other hand, ^{55}Mn, which is stable, and ^{53}Mn, whose half-life is of about $3.7 \cdot 10^6$ years, cannot release a significant dose to the patient. Finally ^{54}Mn, with an intermediate half-life of 312 days, could represent an issue in terms of the expected dose increase to the patient [1]. A possible solution to limit the co-production of these contaminants is represented by the use of highly enriched targets of ^{52}Cr, by means of the reaction ^{52}Cr(p,n)^{52g}Mn, with a concomitant significant cost increase.

In [2], the reaction ^{Nat}V(α,x)^{52g}Mn was identified as a possible alternative route for the production of ^{52g}Mn. Through the use of state-of-art nuclear reaction codes, like TALYS [3], FLUKA [4] and EMPIRE [5], this reaction was shown to lead to a favourable yield of ^{52g}Mn, compared with the standard reaction with ^{Nat}Cr, in combination with a significantly lower production of ^{54}Mn. Nevertheless, the study pointed out that the experimental data available in the literature on the cross section of that reaction are very scattered, and that the standard simulations with the considered nuclear reaction codes cannot reproduce adequately the experimental data. On the other hand, standard TALYS calculations describe the cross section of the routine reaction ^{Nat}Cr(p,x)^{52g}Mn in agreement with the body of experimental data.

In view of possible applications of ^{52g}Mn to medical imaging, in the present work, in the framework of the METRICS project at INFN-LNL, we optimized some parameters of the TALYS code, in order to obtain a better agreement with the experimental measurements for the cross section of the ^{Nat}V reaction and to be able to obtain reliable predictions about the yield and purity of ^{52g}Mn if produced through this alternative reaction route.

2 Methods

In order to perform theoretical calculations of the cross sections for the production of the medical radionuclide of interest, we used modern nuclear reaction codes: in this work we focused in particular on the TALYS simulation package, which is actively developed and maintained. In TALYS, the evaluated nuclear reaction rates are based on the Hauser-Fesbach model [6] for the equilibrium mechanisms and on four different theoretical frameworks for the pre-equilibrium process. Moreover, there are six possible options for the description of the level density, ranging from the Fermi gas model to microscopic approaches. A total of 24 model combinations are thus available.

In a previous work [2] we decided not to select one particular model for level density and pre-equilibrium, but to deal with the theoretical variability provided by the different models. In particular, we introduced for each energy a Best Theoretical Evaluation (BTE) of the cross section, defined as the average between the first and the third quartile. Nevertheless, none of the 24 TALYS models, as well as the BTE, were able to reproduce adequately the experimental data for the cross section of the reaction ^{Nat}V(α,x)^{52g}Mn.

In this work, with the aim of improving the theoretical description of this cross section, we tuned some parameters of the TALYS code. More specifically, we selected, for the optical model, the so-called JLM (Jeukenne-Lejeune-Mahaux) semi-microscopic nucleon-nucleus spherical optical model potential [7] and, for the nuclear level density, we focused on the microscopic approach denoted as Hartree-Fock Models (HFM), specifically the temperature-dependent Hartree-Fock-Bogoliubov model with the tabulation from Hilaire's combinatorial tables [8] (TALYS option: *ldmodel* 6). The tabulated values can be rescaled by using two parameters (c and p), which modulate the HFM level densities according to the transformation:

$$\rho(E) = \exp(c\sqrt{E-p})\rho_{HFM}(E-p) \quad (1)$$

Parameters c and p represent a normalization and an energy shift, respectively. In the present work, we tuned these two parameters, thus modifying the original HFM level densities, in order to obtain theoretical calculations in agreement with the experimental cross sections, for the involved radionuclides. As a first step we performed a gross search, by varying c and p on a grid with a trial and error approach, until the achievement of a first qualitative agreement between theory and measurements. As a second step, we evaluated a global chi-square of all the involved cross sections and we performed a Minuit-based chi-square minimization procedure [9], by using c and p as free parameters, with initial values identified by the previous grid search. These two steps are described in detail in [10]. The computation of reliable cross sections allowed us to identify a promising energy window for the production of the radionuclide of interest with high yield and purity.

We also developed a tool, described in [2], to simulate an irradiation in this energy window and to evaluate the production yield of the radionuclide of interest and of its contaminants, as well as the Isotopic and Radio-Nuclidic Purities (IP and RNP), defined for this specific case respectively as:

$$IP = \frac{n_{^{52g}Mn}}{\sum_i n_{\ ^{i}Mn}} \quad (2)$$

$$RNP = \frac{A_{^{52g}Mn}}{\sum_i A_{\ ^{i}Mn}} \quad (3)$$

where n denotes the number of nuclei and A denotes the activity. These quantities are calculated by computing the production rate of each radionuclide and by solving standard Bateman equations for the radioactive isotopes, as described in [2].

3 Results

Focusing on the cross section of the reaction $^{Nat}V(\alpha,x)^{52g}Mn$, we found that the values c=0.4 and p=-1.0 for the level density of ^{52}Mn led to a minimization of the chi-square, as described in the Methods Section. Since the experimental data are extremely scattered and the different sets of data are not in agreement one each other within their uncertainties, the reduced chi-square assumes a very large value. Nonetheless, the chi-square minimization procedure led us to obtain a good overall agreement between the calculated cross section and the body of the experimental data [11-20]. This agreement is much better than for the BTE, calculated in [2]. The experimental data, the cross section obtained with the optimized c and p values and the BTE are reported in Fig. 1.

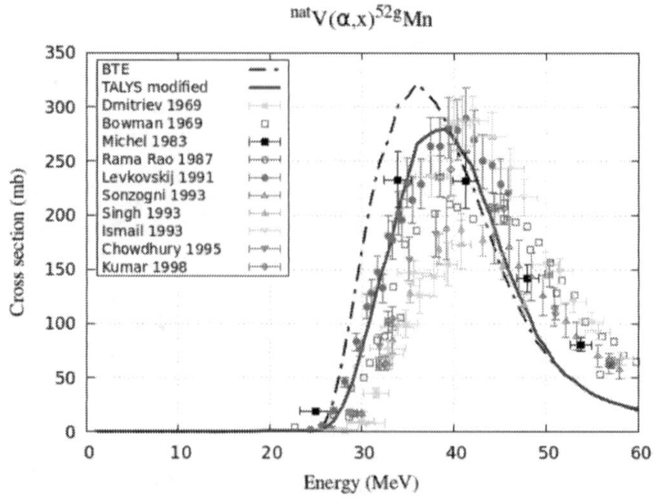

Fig. 1. Experimental data for the cross section of the reaction $^{Nat}V(\alpha,x)^{52g}Mn$, together with the optimized TALYS calculation (indicated TALYS modified) and with the BTE.

This same parameter set led also to a good agreement with the experimental data [15-17, 19-24] for the cross section of ^{54}Mn (Fig. 2), without the need to modify the default c and p values for this radionuclide. A correct description of the cross section of ^{54}Mn is fundamental, since it represents the main contaminant of ^{52g}Mn in the context of medical applications, due to its long half-life (about 312 days).

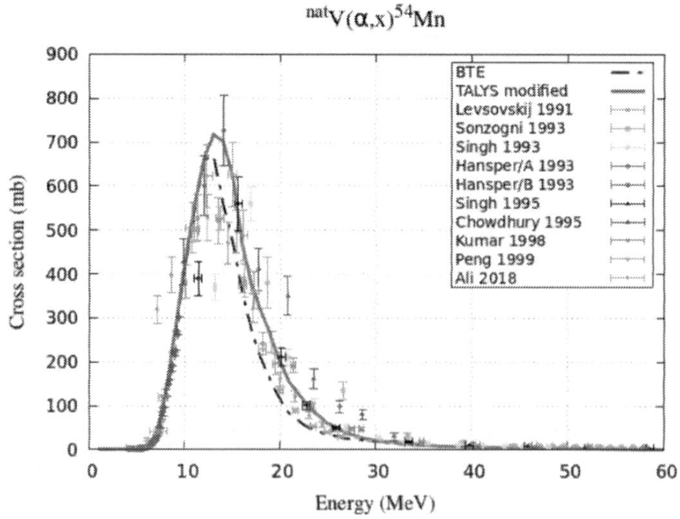

Fig. 2. Experimental and theoretical (BTE and TALYS modified) cross sections for the reaction $^{Nat}V(\alpha,x)^{54}Mn$.

After obtaining reliable theoretical cross sections, in agreement with the experimental data, for both ^{52g}Mn and ^{54}Mn, it is possible to identify a promising energy window to maximize the production of the radionuclide of interest, ^{52g}Mn, and to minimize the production of its main contaminant, ^{54}Mn, as shown in Fig. 3. More specifically, we considered the ratio between the cross section of ^{52g}Mn and the cross sections of all produced Mn isotopes: this ratio reaches its maximum at about 41 MeV. The width of the energy window should correspond to a realistic target thickness: in this case we chose a thickness of 200 μm, as done in [2]. This energy window thus falls between about 34 and 48 MeV.

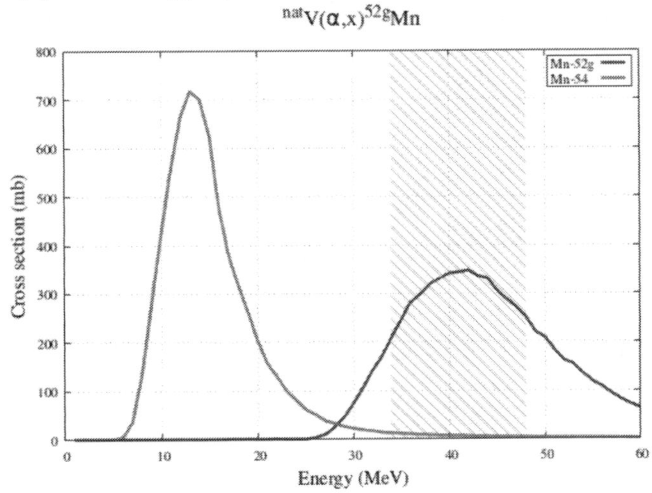

Fig. 3. Comparison between the optimized cross sections of ^{52g}Mn and ^{54}Mn, and identification of a promising energy window for the production of high purity ^{52g}Mn

In order to calculate the expected yields of the radionuclide of interest, ^{52g}Mn, and of its main contaminant, ^{54}Mn, we simulated an irradiation with beam current of 1 μA, irradiation time of 1 h, incident energy of 48 MeV and target thickness of 200 μm (corresponding to an output energy of 33.9 MeV). The results in terms of thick target yield (MBq/μAh) are reported in Table 1.

To compare these results with the standard production route, we also considered the reaction with protons on ^{Nat}Cr, for which the experimental cross section for the production of ^{52g}Mn can be adequately reproduced by standard TALYS calculations, as reported in [2]. Also for the case of ^{Nat}Cr we simulated an irradiation with the same conditions: beam current of 1 μA, irradiation time of 1 h and target thickness of 200 μm. In order to optimize the production of ^{52g}Mn also for this specific case, the incident energy of the protons was equal to 17 MeV, corresponding to an output energy of 14 MeV. The results in terms of physical thick target yield are reported in Table 1, in order to allow a direct comparison with the alternative reaction with ^{Nat}V.

Interestingly, the reaction with ^{Nat}V is expected to lead to a higher yield of ^{52g}Mn than the routine reaction with ^{Nat}Cr, with an associated lower production of the main contaminant ^{54}Mn. These results corroborate the idea of considering the $^{Nat}V(\alpha,x)^{52g}Mn$ reaction as a promising alternative route for the production of ^{52g}Mn.

Table 1. Physical thick target yields for both ^{52g}Mn and ^{54}Mn, produced by the reactions on ^{Nat}V and on ^{Nat}Cr.

Reaction	Physical thick target yield (MBq/μAh)
$^{Nat}V(\alpha,x)^{52g}Mn$	5.16 ± 1.04
$^{Nat}Cr(p,x)^{52g}Mn$	4.40 ± 0.51
$^{Nat}V(\alpha,x)^{54}Mn$	2.71 ± 0.22
$^{Nat}Cr(p,x)^{54}Mn$	4.89 ± 0.07

This idea is also strengthened by the expected RNP for the two considered reactions. Indeed, we determined the time evolution of the produced activities for both the nuclide of interest and all its contaminants, and we calculated the RNP, which was found to be higher for the ^{Nat}V reaction than for the ^{Nat}Cr one, as shown in Fig. 4. It is important to point out that the RNP for ^{Nat}V may be affected by the uncertainties related to the c and p values obtained for ^{52g}Mn cross section. However, the curve presented in Fig. 4 is a reasonable estimate, since it is deduced from cross sections in agreement with the experimental data. In any case, in order to obtain a more accurate RNP, it would be essential to have more precise experimental data for a more accurate calibration of the model parameters.

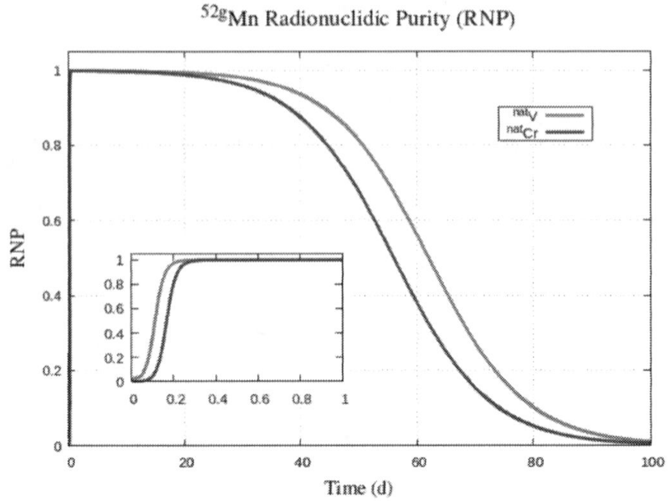

Fig. 4. Comparison between the Radionuclidic Purities of ^{52g}Mn for the two reactions with ^{Nat}V and ^{Nat}Cr, for 1 μA beam current, 1 h irradiation time and 200 μm target thickness. The inset shows the evolution for the first 24 h, including the irradiation time.

4 Conclusions

In the present work we considered the reaction $^{Nat}V(\alpha,x)^{52g}Mn$, as an alternative route to the standard reaction with protons on ^{Nat}Cr, to produce ^{52g}Mn with high purity in view of MultiModal Imaging applications. As a first step we tuned the parameters governing the level density description in a microscopic model implemented in the TALYS nuclear reaction code, in order to find a good agreement between theoretical calculations and experimental data available in the literature. This procedure was performed for both the radionuclide of interest, ^{52g}Mn, and its contaminants, mainly focusing on ^{54}Mn. A promising energy window to maximize the production of ^{52g}Mn with respect to the contaminants was thus identified by analysing the cross sections, and a virtual irradiation was simulated with an appropriate beam energy. The results of this calculation show that the reaction with ^{Nat}V is expected to lead to a higher yield of ^{52g}Mn compared to the standard ^{Nat}Cr reaction, with an associated lower production of the main contaminant, ^{54}Mn. The RadioNuclidic Purity of ^{52g}Mn is also expected to be higher for the ^{Nat}V reaction. These calculations support the idea of considering the $^{Nat}V(\alpha,x)^{52g}Mn$ reaction as a promising alternative route for the production of ^{52g}Mn for medical applications. Nevertheless, it is important to observe that, whereas the production with ^{Nat}Cr can be achieved with hospital cyclotrons exploiting low energy protons, the ^{Nat}V reaction requires 40-MeV alpha particles, only available in advanced research centers. For these centers, however, the new production route may be considered as a valuable alternative, also considering that the half-life of ^{52g}Mn would allow for a relatively long-distance redistribution.

References

1. M. Brandt, J. Cardinale, I. Rausch, T.L. Mindt, J. Label Compd. Radiopharm. **62**, 8 (2019)
2. A. Colombi, M.P. Carante, F. Barbaro, L. Canton, A. Fontana, Nucl. Technol., DOI: 10.1080/00295450.2021.1947122 (2021)
3. S. Goriely, S. Hilaire, A.J. Koning, Astron. Astrophys. **487**, 2 (2008)
4. Battistoni et al., AIP Conf. Proc. **896**, 31 (2006)
5. M. Herman et al., Nucl. Data Sheets, **108**, 12 (2007)
6. W. Hauser, H. Feshbach, Phys. Rev., **87**, 2 (1952)
7. E. Bauge, J.P. Delaroche, M. Girod, Phys. Rev. C, **63** (2001)
8. S. Hilaire, M. Girod, S. Goriely, A.J. Koning, Phys. Rev. C, **86** (2012)
9. F. James, M. Roos, Comput. Phys. Commun., **10**, 343 (1975)
10. F. Barbaro, L. Canton, M.P. Carante, A. Colombi, L. De Dominicis, A. Fontana, F. Haddad, L. Mou, G. Pupillo, Phys. Rev. C, **104**, 044619 (2021)
11. P.P. Dimitriev, I.O. Constantinov, N.N. Krasnov, At. Energy, **26**, 5 (1969)
12. W.W. Bowman, M. BLANN, Nucl. Phys. A, **131**, 3 (1969)
13. R. Michel, G. Brinkmann, R. Stück, Nucl. Data Sci. Tech (Ed. Springer, Dordrecht, 1983)
14. J.R. Rao et al., J. Phys. G., **13**, 4 (1987)
15. V.N. Levkovskij, Cross-Section of Medium Mass Nuclide Activation (A = 40–100) by Medium Energy Protons and AlphaParticles (E = 10–50 MeV), Intervesi, Moskow, Russia, (1991)
16. A.A. Sonzogni et al., J. Radioanal. Nucl. Chem., **170**, 1 (1993)
17. N.L. Singh, S. Agarwal, J.R. Rao, Can. J. Phys., **71**, 3-4 (1993)
18. M. Ismail, Pramana, **40**, 3 (1993)
19. D.P. Chowdhury et al., Nucl. Instrum. Meth. Phys. Res. B, **103**, 3 (1995)
20. B. Bindu Kumar, S. Mukherjee, N.L. Singh, Phys. Scr., **57**, 2 (1998)
21. V.Y. Hansper et al., Nucl. Phys. A, **551**, 1 (1993)
22. N.L. Singh et al., J. Phys. G, **21**, 3 (1995)
23. X. Peng, F. He, X. Long, Nucl. Instrum. Meth. Phys. Res. B, **152**, 4 (1999)
24. B. Ali et al., Pramana – J. Phys., **90**, 3 (2018)

The FOOT experiment: current status and future perspective for nuclear fragmentation studies in particle therapy and space radiation protection

Sofia Colombi[1*], *on behalf of the FOOT collaboration*

[1]Istituto Nazionale di Fisica Nucleare (INFN), Section of Bologna, Bologna, Italy

Abstract. Nuclear fragmentation processes occurring in the beam-target interaction represent a potential source of hazard for humans' health both in particle therapy and space exploration. The FragmentatiOn Of Target (FOOT) experiment aims to characterize both beam and target fragments, and to measure with great accuracy nuclear fragmentation cross sections interesting in cancer therapy and in long-lasting interplanetary missions. In this paper, the fragments reconstruction capabilities of the FOOT experimental setup is reported. FOOT future perspective of providing useful data to improve the modelling of neutron interactions is also discussed.

1 Introduction

The two most relevant sources of radiation in the Solar System are Galactic Cosmic Rays (GCR) and Solar Particle Events (SPE). GCRs energy spectrum has a broad peak between 0.4 and 1 GeV/u, while the typical SPEs energies are below 150 MeV/u. The space radiation field, mainly composed of protons [1], is further modified by interacting with any material composing the spacecraft, including the passive shielding, which is nowadays considered the most effective countermeasure for mitigating radiation exposure. Recent studies [2] questioned the radioprotection advantage of using thicker passive shielding, showing that the build-up of neutrons actually increases the equivalent dose. The possible dominant role of secondary neutrons has a strong impact on the mission risk assessment and can potentially restrict the boundaries of space exploration. Improving the knowledge of nuclear processes relevant for space radiation protection perfectly matches the need in particle therapy. The latter employes protons and carbons beams at energies in the range [60-400] MeV/u to deliver a uniform dose in the tumor region, minimizing the damage to the surrounding healthy tissues. The characterization of both neutral and charged secondary particles produced in the nuclear patient-beam interaction has to be considered in the Treatment Planning System (TPS), in order to account also for the unwanted dose at the target volume edges and far away from the target region [3].

* Corresponding author: colombi@bo.infn.it

2 The FOOT experiment

The FOOT experiment has been designed to identify all the charged fragments produced in the primary beam collision with different targets, aiming to measure both projectile and target fragmentation. The latter is a challenging task due to the short range of the fragments (in the order of 10-100 μm), which results in a very low probability of escaping the target and be detected. To overcome this issue, FOOT relies on an inverse kinematic approach: fragmentation of different ion beams (like ^{12}C and ^{16}O) at 200-1000 MeV/n in hydrogen enriched targets will be studied. Therefore, the proton-nucleus cross sections is extracted by subtraction from data taken on C_2H_4 and C targets [4]. FOOT can also perform the direct measurement of projectile fragmentation cross sections induced by the beam on any targets, including for example dedicated materials for space radioprotection (e.g. Aluminum). FOOT includes a double experimental setup, exploring different angular regions of fragments emission. Below 10 degrees, an Electronic Setup (ES) is employed, while an Emulsion Cloud Chamber (ECC) is used for larger angles. Specifically, the ES consists in a pre-target monitor region to provide the trigger to the whole experiment and the start time for the Time-Of-Flight evaluation. Downstream of the target, there are three elements: i) a magnetic spectrometer for evaluating the production vertex and momentum (p) of the fragments through the tracking outside the magnetic field, ii) a scintillator detector to provide the stop signal for the Time-Of-Flight evaluation and the measurement of energy loss (ΔE) of the fragments, and iii) a calorimeter to measure the kinetic energy (E_{kin}) of all particles. In the following, the acronym *TOF* will refer to the time measured between the above mentioned start and stop signals. A detailed characterization of the ECC and all the detectors involved in the ES can be found in [5].

3 Present performances results

In order to study the expected reconstruction capability of the ES, a study based on experimental-like data samples generated by Monte Carlo (MC) Fluka has been conducted. The samples has been produced by applying the present experimentally and analytically evaluated resolutions on *TOF*, p, ΔE and E_{kin} (i.e., $\sigma(TOF) \simeq 70$ ps, $\sigma(p)/p \simeq 3.7\%$ and $\sigma(E_{\text{kin}})/E_{\text{kin}} \simeq 1.5\%$) as a Gaussian smearing to the corresponding values predicted by MC. The produced fragment charge Z is then evaluated by combining the reconstructed ΔE and *TOF* measurements in the Bethe-Bloch equation [6, 7]. The results show that the precision on the charge reconstruction ranges between ~6% for hydrogen to ~2% for oxygen, allowing a clear identification of the particles produced in the nuclear beam-target interaction. Experimental data were collected in 2019 with a partial FOOT electronic setup at the GSI Helmholtz Center for Heavy Ion Research (Germany) with a 400 MeV/u ^{16}O beam impinging on a 5 mm graphite target (ρ=1.83 g/cm^3). The charge separation obtained on this real data sample is satisfactory and in agreement with the prediction of the MC-based study, thus confirming the goodness of the reconstruction method [8, 9]. The isotopic discrimination is obtained with a standard χ^2 minimization approach by performing a fit to the three reconstructed number of mass values. The latter were evaluated by combining *TOF*, p and E_{kin} measurements in different equations, as follows:

$$A_1 = p/(u\gamma\beta c) \qquad A_2 = E_{\text{kin}}/[(\gamma-1)uc^2] \qquad A_3 = (p^2c^2 - E_{\text{kin}}{}^2)/(2uc^2E_{\text{kin}}) \qquad (1)$$

where u is the Unified Atomic Mass (~931.5 MeV), γ is the Lorentz factor, β is the fragment velocity provided by the tracking path (L) coupled with the *TOF* measurement as $\beta = L/(TOF \cdot c)$. An additional χ^2<5 cut allows to discard the fragments whose kinetic energy has been underestimated, without reducing the statistics more than ~20%. Only A_1 provides a reliable number of mass estimation at higher primary beam energy (i.e., from 700 MeV/u), due to the greater probability of secondary fragmentation and neutrons emission inside the calorimeter, that corrupt the energy measurement. The method ensures a precision on the number of mass reconstruction ranging between ~3.5% and ~4.5%, for all the primary beam energies.

The reconstruction of the charge Z and number of mass A allows to uniquely identify the fth fragment generated in the beam-target interaction, therefore evaluating the corresponding yield Y_f. However, Y_f requires a *background* (Bkg) correction and the application of an *unfolding* (U) procedure [10] to account for the Z and A misidentification probabilities and the possible wrong counts of the produced fragments due to experimental effects. Afterwards, the differential cross-sections (CS) of such f^{th} fragment with respect to its production kinetic energy E_{kin} is retrieved as follow:

$$\sigma_f/dE_{kin} = (Y_f(E_{kin}) - Bgk_f)^U/(N_{prim}N_t\Omega_E\epsilon) \qquad (2)$$

where N_{prim} is the number of primary beam particles, N_t is the number of target scattering center, $\Omega_{E_{kin}}$ is the energy phase space, is the reconstruction efficiency. The total cross section is evaluated by integrating over the whole energy range. The procedure explained for the CS measurements has been applied to data samples produced with both C_2H_4 and C targets, in order to evaluate the result on a H target by difference. For all the targets, the results on the total cross section are compatible within the uncertainties with the MC predictions.

The contribution of each detector on the precision of the mass number determination has been evaluated by varying the resolutions on TOF, p and E_{kin} within a reasonable range around the current real values. The findings point out a strong dependence on the TOF precision, which can worsen the resolution on the number of mass determination of ~0.4%/ps.

4 Future perspectives for neutrons detection

The increasing interest in more precise measurements of neutrons production has led the FOOT collaboration to investigate the possibility to characterize secondary neutrons generated by several projectiles-energy-targets combinations relevant for both particle therapy and space radioprotection. The present ES will allow to identify neutrons with the anticoincidence technique applied to the signals of the last two elements of the apparatus, namely a 6 mm thin plastic scintillator called ToF-Wall (TW) and a BGO Calorimeter (CAL) [5]. An uncharged particle impinging on the CAL will deposit either a very small or null amount of energy in the TW, due to the extremely low interaction probability. Therefore, the apparatus could identify neutrons in the CAL, and determine their kinetic energy with the TOF measurement. To estimate the amount of neutrons arriving in the CAL for detection but not produced in the target, a Fluka MC simulation of a 200 MeV/u ^{16}O beam impinging into a 5 mm C_2H_4 target has been studied. The results show that neutrons are largely produced in the setup detectors, as illustrated in Figure 1. In details, the main sources of neutrons are the permanent magnets and the calorimeter, that produce about two and four times the amount of neutrons originated in the target, respectively. However, only ~20% of the overall neutrons generated will actually reach the CAL, due to the angular acceptance of the detector itself and the probability to interact with other elements of the apparatus or the walls of the irradiation room. Of those arriving in the CAL, the amount of neutrons generated inside the CAL itself is about eight times the production in the target, thus representing the main contribution to the background neutrons. However, the mean kinetic energy of neutrons coming from the target is almost a factor four times higher with respect to the neutrons generated in the CAL. Therefore, a convenient detection energy threshold (e.g., 50 MeV) will help to clean up the neutral particle spectrum from the background contribution, thus selecting only the high-energy neutrons originating from target. A further option currently under investigation regards the possibility to add one or more liquid scintillators specifically designed for the detection of fast (i.e., >1 MeV) neutrons due to the excellent pulse shape discrimination properties [11]. Thanks to their compactness, they will be easy to move to different angles

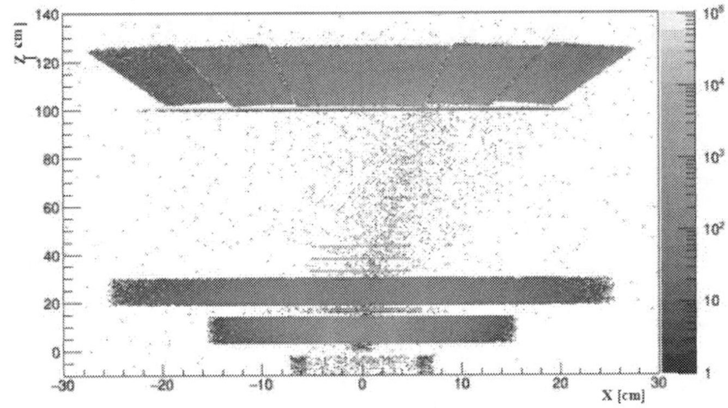

Figure 1. 2-D plot representing the production point of all the neutrons generated in the ES. Z and X are the direction parallel and traversal to the beam direction (in cm), respectively. All the detectors composing the setup presented in Sec.4 are evident. The principal sources of neutrons (i.e., the target placed at -0.25<Z<0.25., the magnets placed at 5<Z<29.5 and the calorimeter placed at 102<Z<126) are clearly visible.

with respect to the primary beam position. Therefore, neutrons production at several angles will be investigated as required for the evaluation of differential cross sections with respect to the emission angle.

5 Conclusions

The FOOT experiment will contribute to increase the knowledge of nuclear processes of charged particles and neutrons production at intermediate energies (200-1000 MeV/u), providing precious experimental data for risk assessment in future space missions and in the development of particle therapy. MC studies suggest a precision in the identification of the produced fragments good enough to accurately reconstruct the differential cross sections with respect to their production kinetic energy. The analysis of the first experimental data collected validates the reconstruction method. The experimental campaigns foreseen in the near future will not only provide data to enforce the reliability of the proposed reconstruction method, but will also deeply investigate the possibility to upgrade the experimental setup to study secondary neutrons production.

References

[1] P. M. O'Neill et al., IEEE Transactions on Nuclear Science **57**, 3148–3153 (2010)
[2] T. C. Slaba et al., Life Sciences in Space Research **12**, 1–15 (2017)

[3] M. Hultqvist et al., Phys. Med. Biol. **55**, 6633-53 (2010)
[4] J. Dudouet et al., Phys. Rev. C **88**, 16 (2013)
[5] G. Battistoni et al., Front. Phys. **8**, 555 (2021)
[6] H. Bethe, Ann. Phys. **397**, 325-400 (1930)
[7] F. Bloch, Z. Physik **81**, 363-76 (1933)
[8] A.C. Kraan et al., Nucl. Instrum. Meth. A **1001**, 165206 (2021)
[9] S. Colombi et al., Phys. Scr. **96**, 114013 (2021)
[10] G. D'Agostini, Nucl. Instrum. Meth. A **362**, 487-498 (1995)
[11] C.A. Goulding et al., LANL Technical Note LA-N2TN-87-201 (1987)

Theoretical study of ^{47}Sc production for theranostic applications using proton beams on enriched titanium targets

Francesca Barbaro[1,2*], *Luciano* Canton[1], *Mario Pietro* Carante[2,3], *Alessandro* Colombi[2,3], *Andrea* Fontana[3]

[1] INFN, Sezione di Padova, Padova, Italy
[2] Università di Pavia - Dipartimento di Fisica, Pavia, Italy
[3] INFN, Sezione di Pavia, Pavia, Italy

Abstract. Recently, scandium-47 has attracted attention in the scientific community thanks to its promising features, making it suitable for targeted radiotherapy and theranostic applications, also in combination with the β^+ emitters ^{43}Sc/^{44}Sc. However, in view of possible pre-clinical and clinical studies, finding efficient production routes is still a current research topic. In this work we investigate ^{47}Sc cyclotron production using proton beams on enriched titanium targets. The analysis of the cross sections and yields of both ^{47}Sc ($T_{1/2}$ = 3.35 d) and its main contaminant ^{46}Sc ($T_{1/2}$ = 83.79 d) has been performed with the nuclear reaction code Talys (v.1.95). The experimental data (scarce and relatively old) are compared with model calculations and some discrepancies emerge even after the tuning of parameters defining the nuclear level densities, involved in the compound nucleus formation. The ^{49}Ti case allows a more precise cross sections reproduction, conversely the ^{50}Ti case requires further theoretical investigations. Preliminary yields analysis has been carried out for both ^{47}Sc and ^{46}Sc.

1 Introduction

Radionuclide ^{47}Sc ($T_{1/2}$ = 3.35 d) represents a promising constituent for innovative radiopharmaceutical compounds for theranostics. Its potential stems from its favorable $\beta-$ decay properties accompanied by the emission of γ-rays of an ideal energy for SPECT imaging. Moreover, it may be easily combined with positron emitters ^{43}Sc ($T_{1/2}$ = 3.9 h) and ^{44}Sc ($T_{1/2}$ = 4.0 h) for production of PET/$\beta-$ theranostics [1].

* Corresponding author: francesca.barbaro@pd.infn.it

Indeed, chemically identical radiopharmaceuticals with the same pharmacokinetics are optimal tools in advanced nuclear medicine. Currently preclinical studies have revealed tumor growth delay and an increased survival time of mice treated with the novel ^{47}Sc DOTA-folate conjugate (cm10), demonstrating the therapeutic potential of this radionuclide. For all these reasons ^{47}Sc is a promising new candidate and a potential competitor of ^{177}Lu for targeted radionuclide therapy [2]. However an efficient and convenient production route has still to be identified and the research is devoted to find a reliable production route which guarantees a sufficient quantity and purity of the radionuclide.
In this work we investigate the cyclotron proton-beam production of ^{47}Sc using ^{49}Ti (5.41%) and ^{50}Ti (5.18%) targets, the only two titanium isotopes that could provide an efficient ^{47}Sc production. Since no measurements for ^{49}Ti targets are available and the sole data for ^{50}Ti are quite old [3], theoretical studies are essential to identify the best irradiation parameters for maximizing ^{47}Sc production on both targets, while limiting the co-production of the ^{46}Sc contaminant ($T_{1/2}$ = 83.79 d) that could affect the purity of the final product. This study is developed within the REMIX (Research on Emerging Medical radIonuclides from X-sections) project, a collaboration among INFN (Istituto Nazionale Fisica Nucleare) Laboratori Nazionali Legnaro, INFN Padova, INFN Pavia, INFN Milano, INFN Ferrara, University of Padova, University of Pavia, University of Ferrara, Istituto Oncologico Veneto, Arronax Nantes France, and Sacro Cuore Don Calabria Hospital Negrar.

2 Production cross sections

The simulation of the relevant cross sections and preliminary yields has been carried out mainly with the nuclear reaction code Talys (v.1.95) [4]. Talys describes the nuclear reaction mechanisms with a variety of models, in particular with 4 pre-equilibrium models (PE) and 6 level density models (LD), for a total of 24 possible combinations of models. Commonly in literature Talys default (PE2-LD1) and Talys adjusted (PE3-LD5) are reference options, but they are not always the most representative ones. For this reason, instead of selecting a specific combination of models we describe the results statistically, taking into account the large variety of models that can be selected with this tool. We refer to 18 combinations of models (PE 1-3 and LD 1-6). We excluded the PE4 model because, in this mass region, its results turned out to be not stable and not physical. In the figure shown in this work the maximum and the minimum of the selected Talys models are represented by dashed lines and the interquartile band, defined as the difference between the upper (Q3) and lower (Q1) quartile by a grey band. It exhibits graphically the dispersion introduced by the variety of models. We denote as the most representative value the BTE "Best Theoretical Evaluation", the median of this interquartile band, depicted as a solid black line. For a comprehensive explanation of this statistical approach, we refer to [5].
The large discrepancies between models and experimental data that emerges from the plots, in particular Fig. 1 and Fig. 4, and the practical aim of the project oriented to medical applications compelled us to improve the model description. Indeed, a better reproduction of the relevant cross sections would allow to achieve a higher accuracy in the predicted yields, purities, and in the assessment of the absorbed dose to the patients' organs. The new analysis approach refers to the microscopic theory of the nuclear reactions where the thermalisation processes use the nuclear level density based on the Hartree-Fock (HF) method. The original HF microscopic level densities can be rescaled by tuning two parameters, c and p, responsible for a normalization change and an energy

shift of the level density curves, respectively. A detailed description of the method is presented in [6].

Predictions for the cross sections of the reactions $^{49}Ti(p,x)^{47}Sc$ and $^{49}Ti(p,x)^{46}Sc$ are presented in Figs 1 and 2. In particular, these figures show the comparison between the statistical approach (BTE curve) and the results of the optimization performed by tuning the c and p parameters trying to reproduce the experimental data available only for the contaminant ^{46}Sc [7]. The best agreement has been achieved with the values $c = 0.11$ and $p = -0.17$ (red line) for the LD4 model and considering the default option for preequilibrium (PE2). The good agreement obtained for the ^{46}Sc case and the lack of ^{47}Sc data from ^{49}Ti targets led us to consider the selected set of parameters for ^{46}Sc to be reliable also for the ^{47}Sc cross section. Starting from the cross sections we identified the optimal energy window that maximizes the yield of ^{47}Sc while minimizing the co-production of ^{46}Sc. Comparing both curves it is evident that the energy range 32 – 40 MeV could optimize the production with ^{49}Ti targets.

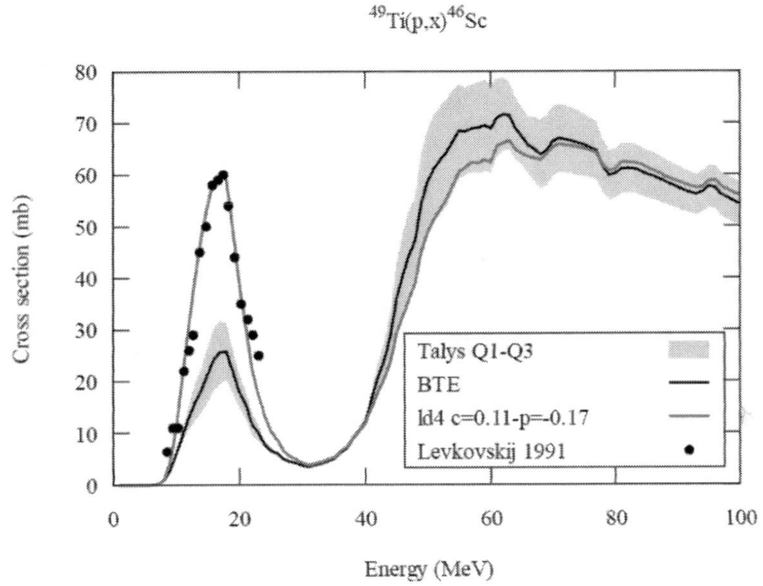

Fig. 1. Experimental and theoretical cross sections for the reaction $^{49}Ti(p,x)^{46}Sc$. The red curve corresponds to the choice of the c and p parameters given in the legend. The grey area is the interquartile band (Q3-Q1) and the black line (BTE) its most representative value.

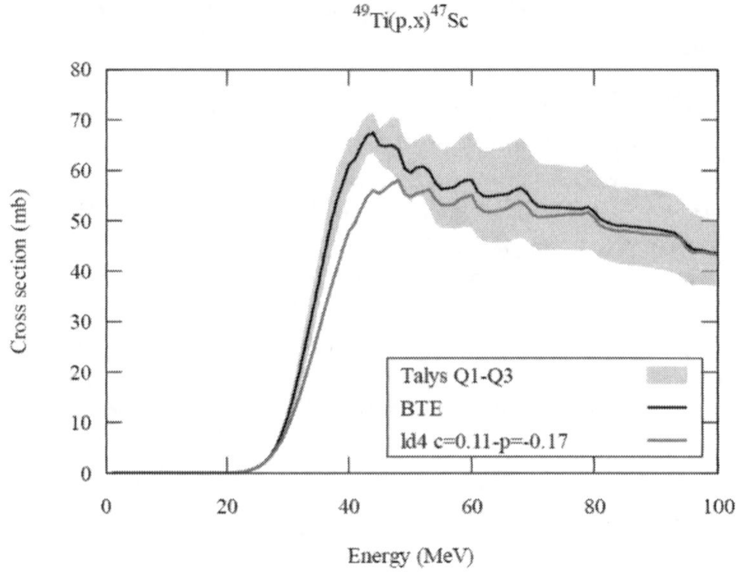

Fig. 2. Calculated cross sections for the reaction ^{49}Ti(p,x)^{47}Sc. No data are available, and the choice of the c and p parameters is based on the tuning performed to reproduce ^{46}Sc data (Fig. 1).

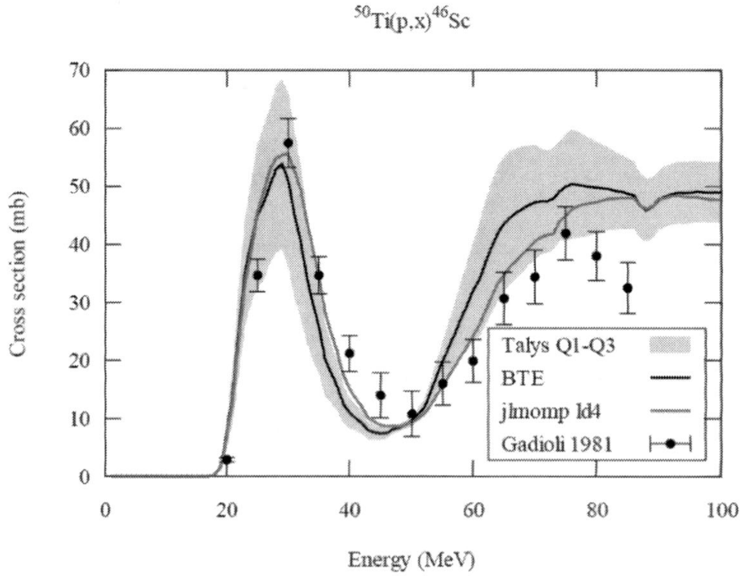

Fig. 3. Experimental and theoretical cross sections for the reaction ^{50}Ti(p,x)^{46}Sc. The Talys modified curve (red line) refers to the ld4 option combined with semi microscopic optical potential JLM. The interquartile band and the BTE curve (black line) refer to a statistical analysis of the standard Talys models.

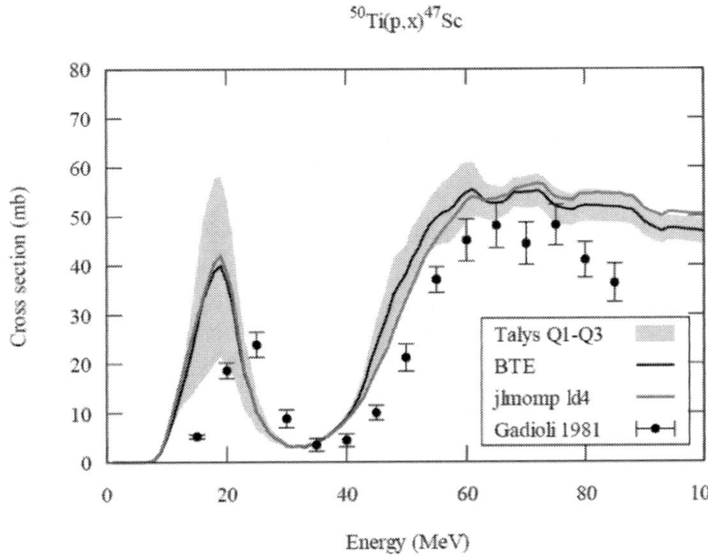

Fig. 4. Experimental and theoretical cross sections for the reaction $^{50}Ti(p,x)^{47}Sc$. Notations and legend same as Fig 3.

The same analysis has been performed for the production routes $^{50}Ti(p,x)^{47}Sc$ and $^{50}Ti(p,x)^{46}Sc$, Figs 3 and 4. For both cases a dataset is available to guide the optimization procedure of the theoretical curves. A good reproduction of ^{46}Sc data is achieved by the combined effects generated by the semi microscopic optical potential JLM (JeukenneLejeune-Mahaux) and by the selected model LD4 without any tuning of the c and p parameters. Conversely for ^{47}Sc cross section from ^{50}Ti all Talys models seem not able to reproduce the data peak around 25 MeV. We noticed the persistence of an offset of about 5 MeV between the data by Gadioli *et al.* [3] and the model results. Further investigations are necessary, and the new experimental data planned at ARRONAX within the REMIX project will be fundamental to validate the theoretical models and to optimize the models free parameters.
In the present situation of uncertainty, we consider the theoretical cross sections to set an optimal range of 8-18 MeV, for ^{50}Ti targets. This range will be employed to evaluate yields, purities, and dosimetric assessments.

3 Yields

In this paper we considered the results obtained within the statistical approach, denoted as BTE cross sections. The analysis performed with the optimized cross section are presently in development. As first step we calculated the production rate for both radionuclides of interest. Secondly, we solved numerically the generalized Bateman equations for the decay chains, whose solutions give the time evolution of the number of nuclei of a certain species produced. Then we evaluated the activity produced by radionuclides and the integral yields at the End of Bombardment (EoB), defined as [8]

$$y_{i,EoB} = \frac{A_i(T)}{I_0} \quad (1)$$

where i indicates the considered radionuclide, $A_i(T)$ is the activity produced by the radionuclide i at the irradiation time T at EoB, and I_0 is the beam current.

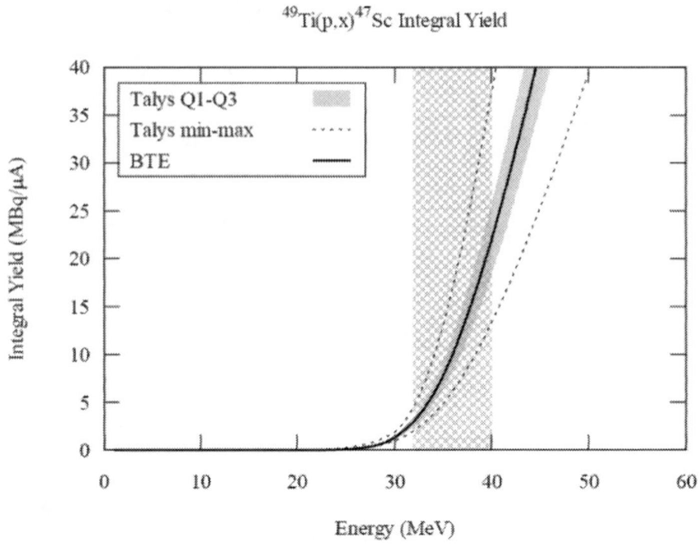

Fig. 5. Predicted yield for ^{47}Sc from ^{49}Ti. The yield has been calculated for a current $I_0 = 1$ μA and irradiation time T = 1h. The green area highlights the energy window (40-32 MeV) that optimizes the production of ^{47}Sc.

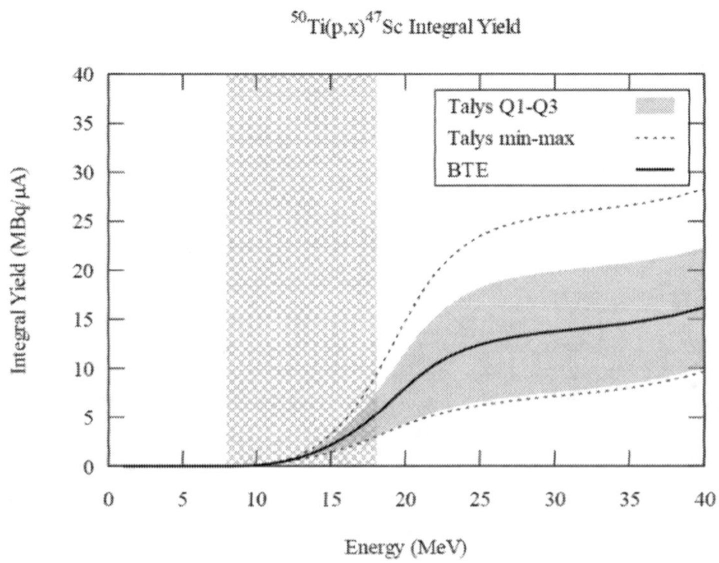

Fig. 6. Predicted yield for ^{47}Sc from ^{50}Ti. The yield has been calculated for a current $I_0 = 1$ μA and irradiation time T = 1h. The green area highlights the energy window (18-8 MeV) that optimizes the production of ^{47}Sc.

The integral yields have been calculated for a current $I_0 = 1$ μA and T = 1 h, with an energy range from E_{max} to 0, namely, assuming full beam energy loss inside the target. Preliminary results of the ^{47}Sc integral yields for both ^{49}Ti and ^{50}Ti cases are plotted in Figs 5 and 6, with the same notation adopted for the other figures (BTE, interquartile band and min/max). We added a green area highlighting the optimal energy window maximizing ^{47}Sc production. From the integral yields plots we obtained the yield for these two specific irradiation energy windows by selecting the values at the corresponding energies E_{in} and E_{out}. The difference $y(E_{in}) - y(E_{out})$ provides the yield for a target of given thickness. In Table 1 we summarize the preliminary yield results for ^{47}Sc and ^{46}Sc from both production routes. The optimized energy windows correspond to a thickness of 1.6 mm for ^{49}Ti target and of 0.93 mm for ^{50}Ti target.

Table 1. ^{47}Sc and ^{46}Sc yields for both ^{49}Ti and ^{50}Ti targets. The irradiation parameters correspond to 1 μA current and 1 h irradiation time.

	$E_{in} - E_{out}$ (MeV)	^{47}Sc Yield (MBq/μA)	^{46}Sc Yield (MBq/μA)
$^{49}Ti(p,x)$	40 – 32	19.00	0.13
$^{50}Ti(p,x)$	18 – 8	5.26	1.08E-4

4 Outlook on dosimetric studies

In view of possible clinical applications of scandium radionuclides, it is important to consider the impact of the radiocompounds to the patients' health. Indeed, to ensure the safety of a radiopharmaceutical and verify the compliance of the European Pharmacopoeia requirements, biodistribution studies and dosimetric calculations are fundamental to assess the impact of the produced radionuclide and main contaminants in terms of dose released to the patient's organs. Contamination is of primary importance, since a radiopharmaceutical must satisfy a minimal purity to be adequate for clinical applications. Nevertheless, it is not sufficient to verify only the percentage of radionuclidic impurities, but it is fundamental to quantify also the dose increase due to the presence of these impurities.

In the near future we plan to develop our study using the CoKiMo (Compartmental Kinetic Model) software [9], that models the human body in compartments connected one to each other and describes the phase of accumulation of the radiopharmaceutical as well as both the fast and slow elimination phases. The approach allows to determine the absorbed dose to a target tissue with the OLINDA (Organ Level Internal Dose Assessment) software, based on the RADAR (RAdiation Dose Assessment Resource) method for internal dose estimation [10]. Finally, combining the OLINDA output with the information of the activity obtained with the irradiation conditions we evaluate the efficacy and quality of the selected radiopharmaceutical.

5 Conclusions

The therapeutic potential of ^{47}Sc has been demonstrated in preclinical trials. However, finding reliable production routes (in terms of quantity and purity) for clinical trials is still an issue that needs to be solved. In this work we report on preliminary results

concerning the ^{47}Sc cyclotron production using proton beams on enriched titanium targets, ^{49}Ti and ^{50}Ti. Cross section results, for both ^{47}Sc and its main contaminant ^{46}Sc, have been obtained with the nuclear reaction code Talys. We have explored the capacity of Talys predefined models to reproduce the data and applied a tuning procedure for the nuclear level densities to improve the cross-section reproduction. We have found that the level density tuning significantly improves the ^{46}Sc production cross section in the case ^{49}Ti targets while no data are available for ^{47}Sc. In the ^{50}Ti case, the ^{46}Sc production cross section is well reproduced by all calculations while for ^{47}Sc the offset of the low-energy peak could not be resolved. New measurements, in progress within the REMIX experiment, will be useful to better clarify the situation with that cross sections.
We calculated preliminarily the BTE yields of both radionuclides for both enriched targets. The results show a higher yield in the case of ^{49}Ti targets, however with ^{50}Ti targets the contamination due to ^{46}Sc is significantly lower. In addition, in this last case, the lower energy range makes the production suitable for typical hospital cyclotrons. These conclusions need to be validated by the planned measurements.
It is expected to complete this study with the evaluation of the dose absorbed in the patients' organs. In fact, in the assessment of the production route, particular attention to contaminants must be paid in order to determine their contribution to the final dose, in line with the European Pharmacopeia requirements.

References

1. C. Müller, K.A. Domnanich, C.A. Umbricht, N.P. van der Meulen, Br. J. Radiol., **91,** 20180074 (2018)
2. C. Müller, M. Bunka, S. Haller, U. Köster, V. Groehn, P. Bernhardt, et al., J. Nucl. Med., **55**, 1658–64 (2014)
3. E. Gadioli et al., Z. Phys. A - Atoms and Nuclei, **301,** 289–300 (1981)
4. S. Goriely, S. Hilaire, A.J. Koning, Astron. Astrophys. **487**, 2 (2008)
5. A. Colombi, M.P. Carante, F. Barbaro, L. Canton, A. Fontana, Nucl. Technol. (2021) https://doi.org/10.1080/00295450.2021.1947122
6. F. Barbaro, L. Canton, M.P. Carante, A. Colombi, L. De Dominicis, A. Fontana, F. Haddad, L. Mou, G. Pupillo, Phys. Rev. C, **104,** 04461 (2021)
7. V.N. Levkovskij, Inter-Vesi: Moscow, Russia (1991)
8. N. Otuka, S. Takács, Radiochimica Acta, **103**, 1-6 (2015)
9. L. Meléndez-Alafort, A. Rosato, G. Ferro-Flores, I. Penev and N. Uzunov, Comptes Rendus 'Academie Bulg. Des ci., **70**, 1649–54 (2017)
10. M. Stabin and J.A. Siegel, J. Nucl. Med., **59**, 154–60 (2017)

New methods for theranostic radioisotope production with solid targets at the Bern medical cyclotron

Gaia Dellepiane[1*], *Pierluigi* Casolaro[1], *Philipp Daniel* Häffner[1], *Isidre* Mateu[1], *Paola* Scampoli[1,2], *Naomi* Voeten[1], *Elnaz* Zyaee[1], and *Saverio* Braccini[1]

[1]Albert Einstein Center for Fundamental Physics (AEC), Laboratory for High Energy Physics (LHEP), University of Bern - Bern, Switzerland

[2]Department of Physics "Ettore Pancini", University of Napoli Federico II, Complesso Universitario di Monte S. Angelo - Napoli, Italy

Abstract. The production of theranostic radionuclides using solid targets is challenging and requires an accurate knowledge of the production crosssections as well as the energy, positioning and focusing of the beam. A research program is ongoing at the 18 MeV Bern medical cyclotron, equipped with a Solid Target Station (STS) and a 6 m Beam Transfer Line (BTL) ending in a separate bunker with independent access. A novel target coin was designed and built to irradiate compressed powder pellets, together with a compact focalization system to optimize the irradiation procedure. Furthermore, methods were developed to measure the beam energy, the production cross-sections and the EoB-activity.

1 Introduction

The concept of theranostics in nuclear medicine couples diagnostic imaging and therapy, using a pair of radioisotopes with identical or very similar chemical properties to label the same biomolecule. A diagnostic radionuclide (β^+- or γ-emitter) is used to detect possible diseases, helping to predict whether the patient will benefit from the treatment imparted by the second radionuclide (β^--, Auger- or α- emitter). Along this line, radiometals can be used to label peptides and proteins and form some of the most promising pairs as $^{43,44}Sc/^{47}Sc$ and $^{61,64}Cu/^{67}Cu$. In contrast to the production of standard radioisotopes with liquid targets, as ^{18}F, the production of radiometals is challenging and requires the use of new irradiation instruments and methods. In particular, rare and expensive isotope enriched materials, usually available in form of powder, have to be irradiated. For this purpose the use of solid target stations represent a possible option. A research program is ongoing at the Bern University Hospital (Inselspital) cyclotron laboratory [1], where an IBA Cyclone 18/18 HC medical cyclotron (18 MeV proton beams, max. 150 μA, 8 out ports) is in operation. The facility is characterized by two bunkers with independent access, allowing to carry out both

* Corresponding author: gaia.dellepiane@lhep.unibe.ch

the routine industrial production of ^{18}F-labeled PET tracers, performed by the spin-off company SWAN Isotopen AG, and multidisciplinary research activities. For the latter purpose, the Bern medical cyclotron is equipped with a 6 m long Beam Transfer Line (BTL) [2], bringing the proton beam to the second bunker, and an IBA Nirta Solid Target Station (STS). To enhance the cyclotron performance, an ultra compact Automatic Focalization System (AFS) based on a novel magnetic lens and a two-dimensional beam detector was recently conceived, constructed and tested [3].

2 Solid target developments and first results

The STS was installed in one of the out ports of the cyclotron and was designed to irradiate a disk (24 mm diameter, 2 mm thick) on which the target material is electroplated. To limit overheating during irradiation, the disk is water-cooled and helium-cooled on the back and the front side, respectively. Since electrodeposition is not suitable for the production of radiometals and other interesting radionuclides, a specific magnetic coin target was conceived and realized by our group. The coin has the same external dimensions as an ordinary disk but it is made of two aluminium or niobium halves kept together by small permanent magnets (Figure 1). To optimize the produced activity and the radionuclidic purity, the material and the thickness of the front-end are used to adjust the energy of protons reaching the target. The back-end hosts the 6 mm pellet and an O-ring to prevent the leakage of molten material or any gas produced during the irradiation.

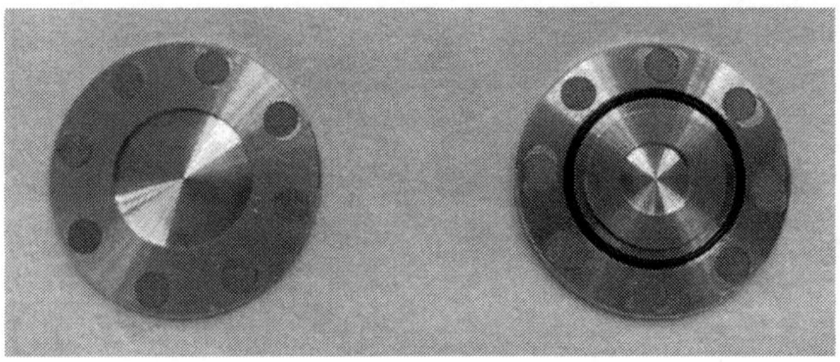

Fig. 1. The two halves of the coin target (24 mm diameter, 2 mm thick).

The STS was complemented with a mechanical transfer system (named Hyperloop - Figure 2-a) [4] designed by our group and with a pneumatic target transfer system (STTS) by TEMA Sinergie (Figure 2-b). The former allows to load the target station without entering the cyclotron bunker, minimizing the dose to the personnel. The latter is used to deliver the irradiated target either to one hot cell in the nearby GMP radio-pharmacy, where the chemical processes are performed, or to a receiving station located in the BTL bunker (Figure 2-c) for further analysis or for transport to external research laboratories. To assess the produced activity, the receiving station was equipped with a CZT detector system based on a ~1 cm^3 CdZnTe crystal (GBS Elektronik). The low counting efficiency makes this detector suitable for the high activities produced; moreover, its position can be modified by means of a programmable step motor, up to a maximum of about 50 cm from the source. Once calibrated by means of a HPGe detector, the signal of the CZT allows assessing the EoB activity with an accuracy of a few percent [5]. Thanks to these developments, several radionuclides have been produced at the Bern medical cyclotron with the STS, as reported in Table 1. In particular, the production of about 15 GBq of ^{44}Sc represents a very promising result in view of theranostic clinical applications [6].

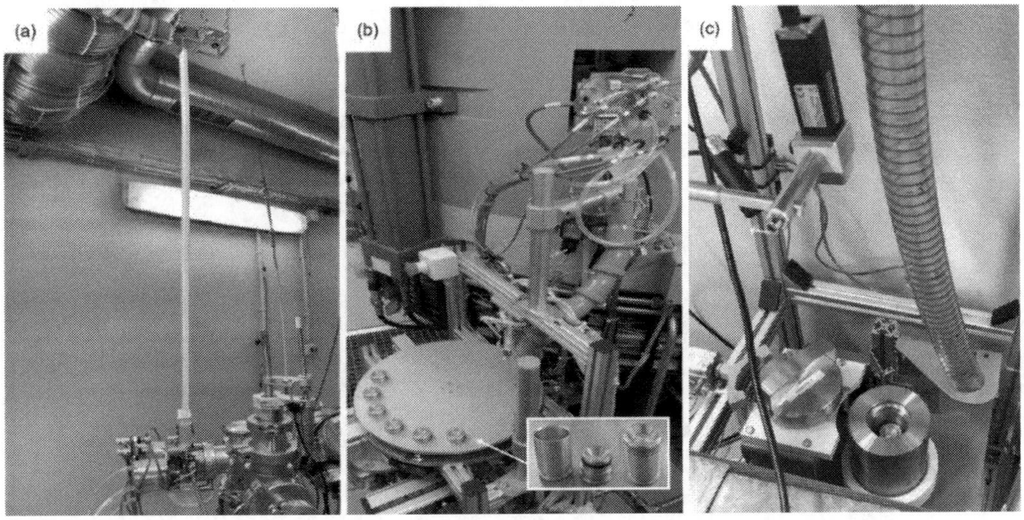

Fig. 2. (a) The Hyperloop system connected to the STS. (b) The IBA Nirta solid target station and the solid target transfer system by TEMA Sinergie. A shuttle is visible in the inset. (c) The receiving station located in the BTL bunker.

Isotope	Reaction	Target	Yield [GBq/μAh]
^{44}Sc	(p,n)	$^{enr44}CaO$	~0.6
^{48}V	(p,n)	^{nat}Ti	~0.015
^{61}Cu	(p,α)	$^{enr64}Zn$	~0.021
^{64}Cu	(p,n)	$^{enr64}Ni$	~0.13
^{68}Ga	(p,n)	$^{enr68}Zn$	~6
^{155}Tb	(p,n)	$^{enr155}Gd_2O_3$	~0.002
^{165}Er	(p,n)	^{nat}Ho	~0.015
^{165}Tm	(p,2n)	$^{enr166}Er_2O_3$	~1.2

Table 1. Main achievements in non-standard radioisotope production obtained with the STS at the Bern medical cyclotron.

3 Automatic Focalization System

In most cases, STSs are mounted in out-ports of medical cyclotrons without any optical device for beam steering and focusing, resulting in an extracted beam of ~12 mm FWHM or more on target. When our coin is used, this implies that about 75% of the extracted protons do not hit the 6 mm target material, producing overheat during the irradiation and unwanted residual activity, giving rise to radiation protection issues.

To limit unnecessary stress for the machine and enhance the irradiation performance, an Automatic Focusing System (AFS) was conceived, developed and recently tested [3]. The AFS consists of the Mini-PET Beamline (MBL) [7], a two-dimensional beam profiler (UniBEaM) and a software feedback system. The MBL is a ~50 cm long magnet produced by the Canadian company D-Pace, embedding two quadrupole and two steering magnets within the same structure. The UniBEaM detector [8] is a non-destructive two-dimensional beam profiler based on scintillating silica doped fibers passing through the beam. This detector was developed by our group to measure the beam profile in a wide range of currents and a commercial version is manufactured by D-Pace [9]. The feedback system analyses the beam profiles measured by the UniBEaM and, if necessary, corrects the beam position and shape by varying the current in the power supplies of the MBL. Beam tests were successfully carried out by installing the AFS at the end of the BTL; in particular we demonstrated the focusing capabilities of the system and a gain factor of about 20 was found in the production of radioisotope using solid targets [3]. The AFS was recently installed in front of the STS in the cyclotron bunker (Figure 3) and the first tests are presently ongoing.

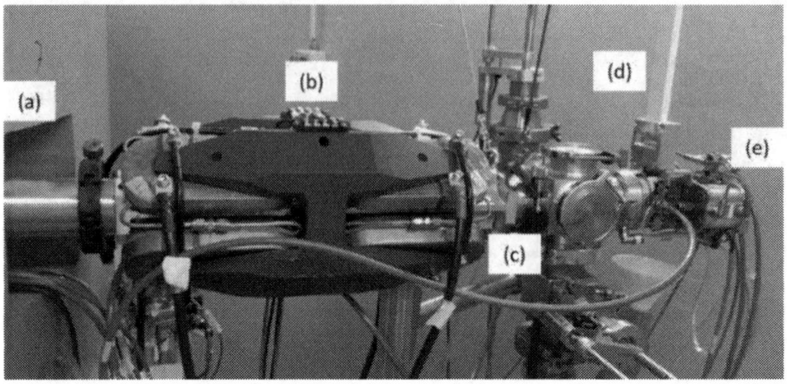

Fig. 3. The Automatic Focalization System installed in the cyclotron bunker: (a) cyclotron out-port; (b) Mini-PET beam line; (c) UniBEaM detector; (d) target receiving station of the Hyperloop; (e) Solid Target Station.

4 Cross section measurements

For an optimized radioisotope production, a precise knowledge of the reaction cross sections is crucial and a novel method for their measurements was developed by our group using targets in form of powder [10, 11]. In most cases, the powder is deposited on an aluminium disk by sedimentation, which means that the target thickness cannot be precisely controlled. For this reason, the new procedure is based on the irradiation of the full target mass with flat proton beams, instead of the usual method based on the irradiation of a homogeneously thick target. The beam is flattened by the optical elements of the BTL and its position and shape are measured on-line with a commercial version of the UniBEaM detector. The beam current hitting the target is measured by means of a custom target station composed of a 6 mm diameter collimator, an electron suppressor ring and a target holder. The collimator is grounded and provides a beam of controlled diameter, within which the beam flatness is better than 5%. The electron suppressor ring is connected to a negative bias voltage in order to repel secondary electrons produced during the irradiation, that would increase the measured beam current. Current measurements on target are performed throughout each irradiation by connecting the target holder to an ammeter (B29885A Keysight). The conductive parts of the target station are kept together by insulator components. The beam energy is degraded by means of aluminium attenuator disks placed in front of the target and is determined using the SRIM Monte Carlo code [12]. The produced activity is assessed by gamma spectroscopy with an HPGe detector.

The target consists of a 23 mm disk, usually made of aluminum, with a 4 mm diameter and 0.8 mm deep pocket in its center (Figure 4-a). The target material is deposited by sedimentation from a suspension of a few milligrams of material in ultra-pure water or ethanol, which is then evaporated by means of a heating plate (Figure 4-b - Figure 4-d). The target mass is assessed with an analytical balance (METTLER TOLEDO AX26 DeltaRange, Figure 4-e). To prevent the escape of the material during the irradiation and the measurement procedure, the disk is covered by a 13 μm thick aluminum foil (Figure 4-f). With this method, the production cross section of several radioisotopes (^{43}Sc, ^{44}Sc [10], ^{47}Sc [11], ^{48}V [10], ^{61}Cu, ^{68}Ga, ^{155}Tb, ^{165}Tm and related impurities) was measured.

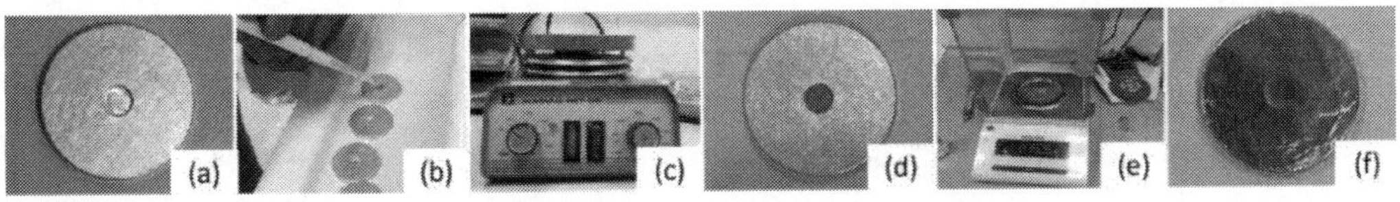

Fig. 4. Preparation of a target for cross section measurements: (a) empty aluminum disk; (b) sedimentation procedure; (c) heating plate; (d) aluminum disk filled with target material; (e) analytical balance; (f) aluminum disk covered with a thin aluminum foil.

5 Beam energy measurement

The energy of the proton beam is another essential parameter for an optimised radioisotope production with solid targets. The energy measurement of the beam extracted on the STS is currently ongoing with a method based on the stacked-foil technique. The well known monitor reaction ^{nat}Ti(p,x)^{48}V was chosen for this purpose, being characterized by a particular shape so that the produced activity strongly depends on the beam energy [13]. Two stacks composed of natural Ti thin foils, separated by a Nb energy degrader, are positioned in the coin pocket and irradiated. Each foil is then measured with a well calibrated HPGe detector, whose counting efficiency is known with an uncertainty less than 5%. Knowing the produced activity, the ^{48}V production cross section can be estimated for several initial proton energies, allowing to determine which energy best reproduces the experimental data. This method has been validated in the BTL, where the extracted beam energy was measured to be (18.3±0.3) MeV, in good agreement with the measurements we performed with other methods [14, 15].

6 Conclusions and outlook

In the framework of the research program on the production of theranostic and non standard radioisotopes ongoing at the Bern medical cyclotron, new instruments and methods were developed. A novel automatic focusing system was conceived, constructed and tested, together with new non-destructive beam monitoring detectors and a novel target coin to irradiate solid materials. New procedures to measure the nuclear cross sections, the beam energy and the produced activity at EoB were also developed.

These developments and findings lead to promising results that will be instrumental in view of the use of novel radioisotopes for theranostics in nuclear medicine.

We acknowledge contributions from LHEP engineering and technical staff (Roger Hänni and Jan Christen in particular). We are thankful to the SWAN Isotopen AG maintenance team (Riccardo Bosi and Michel Eggemann in particular) for the collaboration in the set-up and the operation of the solid target system. This research was partially funded by the Swiss National Science Foundation (SNSF). Grants: 200021_175749 and CRSII5_180352.

References

1. S. Braccini *et al.*, Proceedings of Cyclotrons2019, Cape Town, South Africa, 127-131 (2019)
2. S. Braccini *et al.*, AIP Conf. Proc. **1525** 144-150 (2013)
3. P.D. Häffner *et al.*, Applied Sciences **11**, 2452 (2021)
4. G. Dellepiane *et al.*, Proceedings of the International Conference on Technology and Instrumentation in Particle Physics TIPP2021, Vancouver, Canada, on-line format, (2021)
5. G. Dellepiane *et al.*, IL NUOVO CIMENTO **44 C**, 130 (2021)
6. N.P. van der Meulen *et al.*, Molecules **25**, 4706 (2020)
7. M.P. Dehnel *et al.*, Proceedings of Cyclotron2013, Vancouver, Canada, 251-253 (2013)
8. M. Auger *et al.*, Journal of Instrumentation **11**, P03027 (2016)
9. D.E. Potkins *et al.*, Physics Procedia **90**, 215-222 (2017)
10. T.S. Carzaniga *et al.*, Applied Radiation and Isotopes **129**, 96-102 (2017)
11. T.S. Carzaniga *et al.*, Applied Radiation and Isotopes **145**, 205-208 (2019)
12. J.F. Ziegler and J.M. Manoyan, Nucl. Instrum. Methods B **35**, 215 (1988)
13. S.J.C. do Carmo *et al.*, Instruments **3**, 20 (2019)
14. P.D. Häffner *et al.*, Instruments **3**, 63 (2019)
15. K.P. Nesteruk *et al.*, Journal of Instrumentation **13**, P01011 (2018)

Preliminary results of the experimental cross sections of the long-lived β^+ emitters of interest in PET range verification in proton therapy at clinical energies.

Teresa Rodriguez-Gonzalez[1,2*], *Carlos* Guerrero[1,2**], *Claus Maximilian* Bäcker[3,4,5,6], *Christian* Bäumer[3,4,5,6,7], *Walter* Jentzen[5,6,8], *M. Carmen* Jiménez-Ramos[2], *M. Ángeles* Millán-Callado[1,2], *Beate* Timmermann[4,5,6,7,9], and *José Manuel* Quesada[1]

[1]Dept. de Física Atómica, Molecular y Nuclear. Universidad de Sevilla. Avda. Reina Mercedes s/n, Seville 41012, Spain.

[2]Centro Nacional de Aceleradores (Universidad de Sevilla-Junta de Andalucía-CSIC). C/ Thomas Alva Edison 7, Seville 41092, Spain.

[3]TU Dortmund University, Faculty of Physics, Otto-Hahn-Str. 4a, D-44227 Dortmund, Germany.

[4]West German Proton Therapy Centre Essen, Hufelandstr. 55, D-45147 Essen, Germany.

[5]West German Cancer Center, Hufelandstr. 55, D-45147 Essen, Germany.

[6]University Hospital Essen, Hufelandstr. 55, D-45147 Essen, Germany.

[7]German Cancer Consortium (DKTK), Heidelberg, Germany.

[8]University Hospital Essen, Clinic for Nuclear Medicine, Hufelandstr. 55, D-45147 Essen, Germany.

[9]University Hospital Essen, Department of Particle Therapy, Hufelandstr. 55, D-45147 Essen, Germany.

Abstract. In proton therapy, offline PET range verification requires a comparison of the measured and expected β^+ activity distributions produced by the proton field in the body, looking at the the long-lived β^+ emitters. The reliability of the expected activity distributions depends on the Monte Carlo simulations and hence on the accuracy of the underlying cross section data. However, several studies confirm the need for more and better measurements and evaluations of these cross sections.

In this work, the employed method to measure the production yields of the long-lived β^+ emitters of interest in PET range verification ^{11}C ($t_{1/2}$ = 20 min), ^{13}N ($t_{1/2}$ = 10 min) and ^{15}O ($t_{1/2}$ = 2 min) in C, N and O is presented. The method combines the multi-foil activation technique with the subsequent measurement of the induced activity in a clinical PET scanner. The preliminary results of the^{12}C(p,pn)^{11}C reaction cross sections is presented.

* Corresponding author: mrodriguezg@us.es

** Corresponding author: cguerrero4@us.es

1 Motivation and objective

In comparison to conventional radiation therapy, ion beam therapy is able to reduce the deposited dose in the healthy tissues close to the tumor, due to the characteristic dose distribution of the charged particles in the body of the patient. The proton dose distribution is defined by its maximum dose deposition near the end of the beam range (Bragg peak) and its finite penetration in matter [1]. Despite the mentioned benefits, current treatment plannings have to consider significant safety margins associated to the uncertainties in the actual beam range, due to several physical effects [2].

In this context, Positron Emission Tomography (PET) range verification has been proved to be a well-suited technique to monitor the range of the proton beam. This technique consists in the measurement of the activity distribution induced in the body of the patient with a PET scanner, just after the treatment (using long-lived isotopes, with a half-life of ~min) or during the treatment (using short-lived isotopes, with a half-life of ~ms and ~s) [3]. The use of short-lived isotopes (in-beam range verification) would provide immediate feedback on the dose delivery and avoid the washout effect, which are the main limitations of offline (i.e. delayed) range verification [4] [5]. However, clinically, the implementation of a PET scanner for real-time imaging requires technology and device development, while offline PET range verification can be simply implemented using a conventional PET scanner close to the treatment room.

In addition, PET range verification requires the comparison of the measured activity distribution with a simulated one [6] [4], whose accuracy depends on the accuracy of the underlying cross section data for producing the β^+ emitters of interest. This work is focused on the production of the long-lived isotopes ^{11}C ($t_{1/2}$=20.4 min), ^{13}N ($t_{1/2}$=9.97 min) and ^{15}O ($t_{1/2}$=122 s), of interest in offline PET range verification. The corresponding data currently available show significant discrepancies and indeed several studies [2] [7] confirm the need for more accurate measurements. This work aims to improve the knowledge of the production cross section of the mentioned long-lived β^+ emitters at clinical energies. Furthermore, a parallel experimental campaign is focused on measuring the production cross section of the short-lived isotopes on interest in real-time PET verification.

2 Experimental set-up

The methodology followed for this purpose combines the multi-foil activation technique with a PET scanner [9]. Using a multi-foil activation, it is possible to obtain the production yields at different energies with a single irradiation, as the proton beam energy decreases as it traverses the different foils and hence the activation occurs at a different proton energy in each of them. Then, the simultaneous measurement of the irradiated films by a PET scanner allows to minimize the systematic errors associated to the measurement of the activity.

2.1 Irradiations at the West German Proton Therapy Center (WPE)

The experiment was carried out in one of the clinical treatment rooms at the West German Proton Therapy Center, in Essen (Germany). The number of protons per Monitor Unit (MU) is measured with an ionization chamber. The pencil beam scanning (PBS) mode is used for the activation of the target.

Three multi-foil targets rich in carbon (polyethylene), oxygen (silicon dioxide) and nitrogen (aluminium nitride) were irradiated (see Figure 1, left). The thickness was 1 mm for

https://doi.org/10.1051/epjconf/202226105007

the PE and SiO_2 foils and 0.64 mm for the AlN foils. These thin foils were embedded between thicker foils of polyethylene, which serve to degrade the proton beam energy. The thickness of these degraders is 5 and 10 mm, depending the initial beam energy, to minimize the number of irradiations, thus saving beam time, money and reducing systematic errors . As an example, Figure 2 displays the proton energy distribution for the PE foils calculated with Geant4 [10] for different initial beam energies.

The first foil in each stack is a PMMA *monitor* film, which serves to validate at the same time the measurement of the relative beam intensity and the calculations of the decay during

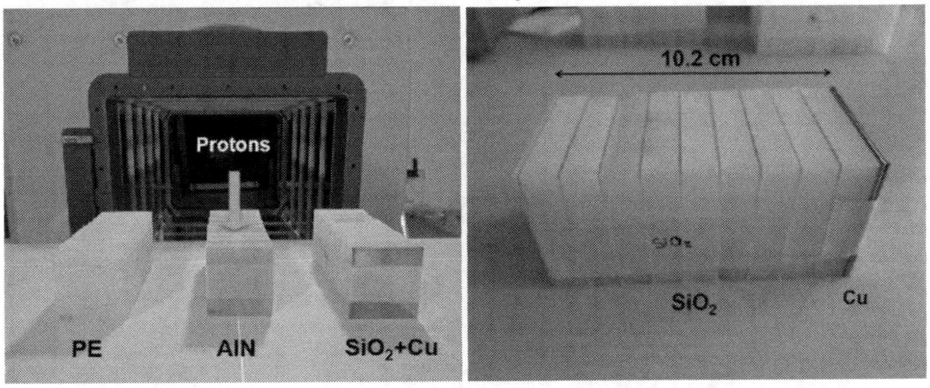

Figure 1. Left: Experimental setup during the irradiations. Right: The SiO_2 stack made of 10 thin foils of SiO_2 between thicker PE degraders. The first and last foils are the PE monitor foil and copper, respectively.

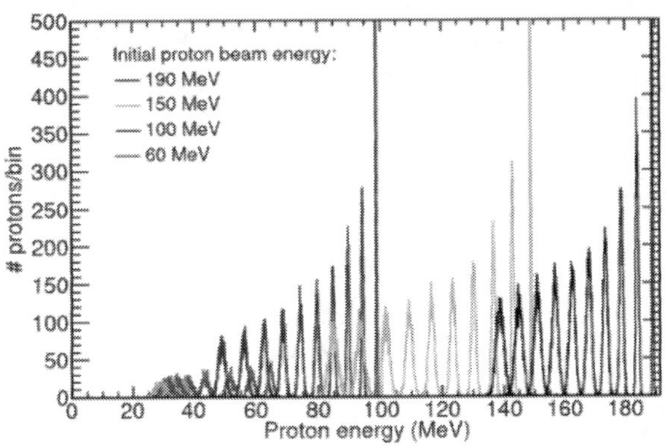

Figure 2. Proton energy distribution for the PE assembly. Note that in the case of the 60 MeV irradiation, a 100 MeV proton beam was degraded using 40 mm of polyethylene. In this case, the different intensities of red colour is used to distinguish between consecutive distributions.

the irradiation and during the transport from the treatment room to the PET scanner. A copper target was also placed at the end of the SiO_2 stack to ensure the validation of the results comparing with the IAEA monitor reaction of reference $^{nat}Cu(p,x)^{63}Zn$ [11]. A picture of the target assembly is shown in Figure 1 (right). Before each irradiation, radiochromic foils EBT3 were placed at the same place than the target foils to ensure the position and size of the beam spot.

2.2 Measurement of the induced activity with a clinical PET/CT scanner

For each initial beam energy, the activity of all foils was measured simultaneously using a PET scanner in dynamic acquisitions of 60 s during 4 hours. In total, 10 foils of PE, 10 foils of AlN, 10 foils of SiO_2, the monitor foils and copper were embedded in thicker plates of polyethylene, to ensure the conversion of the positron emitted into two 511 keV photons. In Figure 3, the polyethylene plates containing the irradiated foils and the PET measurement are shown.

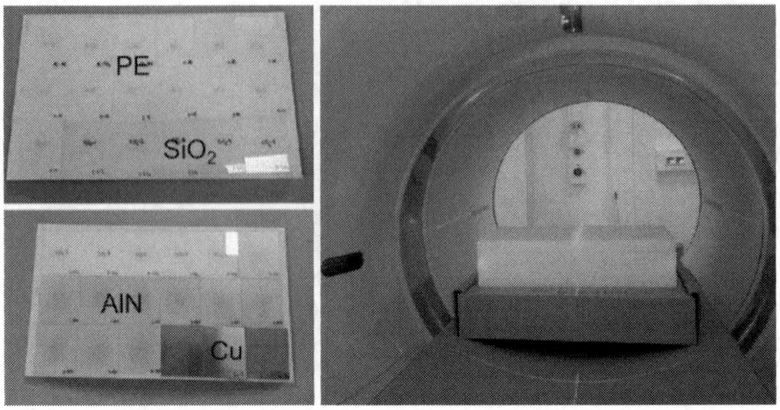

Figure 3. PET/CT measurement of the irradiated foils embedded between polyethylene plates.

The calibration of the PET scanner was performed using a ^{22}Na calibrated source. The uncertainty in the spatial PET efficiency was considered to be 3%, after measuring with sources of different activities in different positions.

3 Analysis and preliminary results of the reaction cross section $^{12}C(p,n)^{11}C$

The reaction cross section was obtained from the induced activity at the end of bombardment (EOB). Taking into account the decay constant of ^{11}C, ^{13}N and ^{15}O, the activity curves obtained with the PET/CT scanner are fitted to:

$$A(t) = c + A^{EOB}_{^{11}C} e^{-\lambda_{^{11}C} t} + A^{EOB}_{^{13}N} e^{-\lambda_{^{13}N} t} + A^{EOB}_{^{15}O} e^{-\lambda_{^{15}O} t}. \quad (1)$$

In equation 1, $\lambda_i = ln(2)/t^i_{1/2}$ and A_i^{EOB} is the activity at EOB, for i each isotope of interest, and c is a constant, that has been proved to be negligible in all cases.

The reaction cross section is then calculated as:

$$\sigma(E_i) = \frac{Y_i(E_i) \cdot c_{decay}}{\lambda \cdot n_i} = \frac{(A^{EOB}/C) \cdot c_{decay}}{\lambda \cdot n_i}, \quad (2)$$

where n_i is the number of nuclei per unity of area, C is the total charge, λ is the decay constant. The induced activity is corrected by the decay during the irradiation process and during the transportation to the PET/CT scanner:

$$c_{decay} = \frac{\lambda T_{irrad}}{1 - e^{-\lambda T_{irrad}}} \tag{3}$$

The analysis of the monitoring foils of polyethylene indicates an agreement between the corresponding A_{EOB} within 1.7% (see Figure 4), which in then considered the uncertainty related to the reproducibility of the irradiations. The good agreement validates also the decay corrections and beam current measurement.

The production cross sections of ^{11}C, ^{13}N and ^{15}O in C, N and O has been measured using this technique. The preliminary cross section of $^{12}C(p,n)^{11}C$ is shown in Figure 5, together with the available data sets in the EXFOR data base [8]. The results presented herein are in

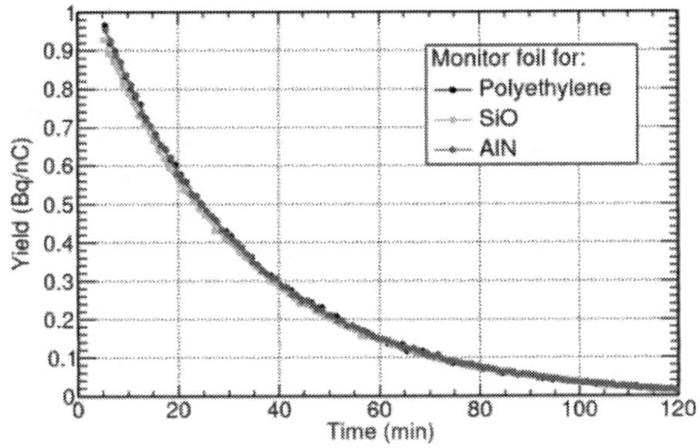

Figure 4. Decay curves of the monitoring foils at 60 MeV.

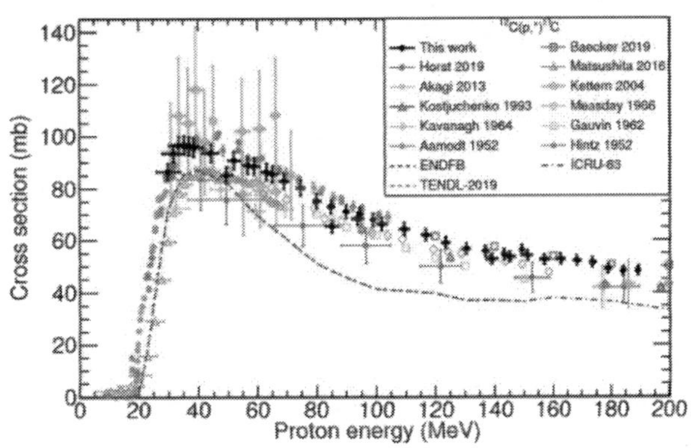

Figure 5. Preliminary production cross section of $^{12}C(p,pn)^{11}C$.

a very good agreement with one of the most recent works, i.e., the experimental data from Bäcker et al. [12].

With this technique, a total of 40 cross section values between 30 and 200 MeV have been obtained. The uncertainty in the cross section values has been calculated taking into account the uncertainty in the PET measurement (3%), the ^{22}Na source (1.5%), the beam current (1%), the foil thickness (1%) and the uncertainty in the fit, obtained with the ROOT software [13] whose algorithms are based on the minimum chi-squared method.

The very good agreement between the overlapped points from two consecutive irradiation energies validates the results and ensures the reproducibility of the measurement.

4 Conclusions and outlook

As a response to the need for more accurate production cross sections of β^+ emitters for PET range verification in proton therapy, this project aims to the measurement of the long-lived isotopes ^{11}C, ^{13}N and ^{15}O in C, N and O. The applied technique provides reliable and accurate measures of the cross sections of interest, measuring several reaction channels in the same experiment, hence minimizing the number of irradiations and systematic errors in the measurement of the activity. The preliminary results of $^{12}C(p,n)^{11}C$ cross section values is presented, together the available data sets in literature. The good agreement between the data sets obtained by different irradiations constitutes a cross-check for the measurement and technique applied herein. The technique itself has been validate using monitor foils and an IAEA cross section of reference. The preliminary results of $^{12}C(p,n)^{11}C$ represent then a validation to obtain the cross section of the reaction channels: $^{16}O(p,*)^{11}C$, $^{16}O(p,*)^{13}N$, $^{16}O(p,*)^{15}O$, $^{14}N(p,*)^{11}C$, $^{14}N(p,*)^{13}N$ and $^{14}N(p,*)^{15}O$. Further analysis and final results of all reaction channels will be submitted for publication.

This project has received funding from the Spanish Ministry of Economy and Competitiveness projects RYC-2014-15271, FPA2016-77689-C2-1-R and RTI2018-098117-B-C21, from the European H2020-847552 (SANDA) and the V Plan Propio de Investigación Programme from the University of Sevilla. T. Rodriguez-Gonzalez acknowledges the Spanish FPI predoctoral grant.

References

[1] A. C. Knopf and A. Lomax, Physics in Medicine and Biology **58** 131-160 (2013)
[2] H. Paganetti et al., Physics in Medicine and Biology **57** 99-117 (2012)
[3] A. C. Kraan et al., Frontiers in Oncology **5** 150 (2015)
[4] K. Parodi et al., Physics in Medicine and Biology **52** 3369-3387 (2007)
[5] X. Zhu and G. El Fakhri, Theranostics **3,10** 731-40 (2013)
[6] U. Oelfke, G. K. Lam and M. S. Atkins, Physics in Medicine and Biology **41** 177 (1996)
[7] S. España et al., Physics in Medicine and Biology **56** 2687 (2011)
[8] EXFOR data base, https://www-nds.iaea.org/exfor/
[9] T. Rodriguez et al., Radiation Physics and Chemistry **190** 109759 (2022)
[10] Allison et al., Nucl. Instrum. Meth. A **835** 186-225 (2016)
[11] Hermanne et al., Nucl. Data Sheets **148** 338-382 (2018)
[12] C. M. Bäcker et al., Nuclear Inst. and Methods in Physics Research B **440** 75–81 (2019)
[13] CERN, "ROOT. Data Analysis Framework". https://www.root.cern.ch

Improvement of nuclear reaction modeling for the production of ^{47}Sc on natural vanadium targets for medical applications

Alessandro Colombi[1,2*], *Francesca* Barbaro[3,1], *Luciano* Canton[3], *Mario Pietro* Carante[1,2], and *Andrea* Fontana[2]

[1]Università di Pavia – Dipartimento di Fisica, via A. Bassi 6, 27100 Pavia, Italy
[2]INFN - Sezione di Pavia, via A. Bassi 6, 27100 Pavia, Italy
[3]INFN - Sezione di Padova, via F. Marzolo 8, 35131 Padova, Italy

Abstract. The proton-induced reaction on natural vanadium targets is studied for the production of the innovative theranostic radionuclide ^{47}Sc as well as of its contaminants, mainly ^{46}Sc. The theoretical excitation functions are calculated using the nuclear reaction code TALYS and are compared with the most recent experimental data. A better agreement between the theoretical curves and the data is achieved with an optimization of the nuclear level density parameters. The obtained improvements represent a useful and important result for accurate evaluations of yields and purities which are needed quantities for subsequent dosimetric studies, in view of the radiopharmaceutical applications of ^{47}Sc. The optimization procedure is explained and shown for ^{47}Sc and ^{46}Sc, and also a comparison among the theoretical and experimental cumulatives is given (for the main contaminant) in addition to an estimation of the production yields for two irradiation conditions for both nuclides.

1 Introduction

In the last few years the production of theranostic radionuclides has gained an increasing interest of the scientific community for their important applications in nuclear medicine. In particular ^{47}Sc represents, at the moment, one the most promising radionuclides to be studied [1, 2] due to its interesting decay characteristics ($T_{1/2}$=3.35 d, E_γ=159 keV, I_γ=68.4%, $E_{\beta-}$=162 keV) [3] which, on one side, make it suitable to study the slow biodistribution of large molecules and to perform SPECT (Single Photon Emission Computed Tomography) imaging, and on the other side to treat small size tumours thanks to the high intensity β− decay. Furthermore, the combination of ^{47}Sc with other Sc-isotopes as ^{43}Sc or ^{44}Sc, which are positron emitters, is very interesting for the possibility to integrate the diagnostic PET (Positron Emission Tomography) studies [2, 4] to the therapeutic applications of ^{47}Sc.

* Corresponding author: alessandro.colombi@pv.infn.it

In this field, nuclear physics studies play a great role especially to select interesting production routes and the most promising conditions to produce the nuclide of interest with sufficiently high yield and purity for its nuclear medicine applications [5]. Indeed, particular emphasis is devoted to the evaluation of the main contaminants, mainly ^{46}Sc ($T_{1/2}$ = 83.79 d), which can affect the final purity and contribute to the final dose imparted to the patients. At the moment all the possible reaction routes are investigated to produce ^{47}Sc with both accelerators and nuclear reactors. In particular, this work is focused on proton-induced nuclear reactions using natural vanadium targets, which have been studied within the PASTA project (Production with Accelerators of Sc-47 for Theranostic Applications) in the framework of LARAMED (LAboratory of RAdioisotopes for MEDicine) program at INFNLegnaro National Laboratories (INFN-LNL), with the final goal to produce the nuclide of interest by using the INFN-LNL 70 MeV proton cyclotron. Nevertheless, at the moment, also other reactions are under study using titanium targets [6].

As described in Ref. [7], for the $^{nat}V(p, x)^{47}Sc$ reaction the theoretical excitation functions, calculated using the nuclear reaction code TALYS [8], are not in agreement with the available experimental data, representing a problem in the evaluation of ^{47}Sc and of its contaminants in view of the theranostic applications. Therefore, in this work our aim is to find a better description of the cross sections investigating the most recent microscopic models implemented inside the TALYS code. Our attention is focused on the nuclear level density parameters to reproduce especially the cross sections of ^{47}Sc and ^{46}Sc, and also the cross sections of all the other nuclides studied in the project (^{43}Sc, ^{44g}Sc, ^{44m}Sc, ^{48}Sc, ^{42}K, ^{43}K, ^{48}V, ^{48}Cr, ^{49}Cr, ^{51}Cr).

2 Nuclear reaction modeling

In the field of radiopharmaceutical production, the use of nuclear reaction codes is essential to model a nuclear reaction and to estimate the yields of the nuclide of interest as well as of the contaminants, starting from the studies of the cross sections. Nuclear reactions are very complex to analyze since several aspects have to be taken into account, as the co-existence of the different reaction mechanisms, like the compound nucleus formation and evaporation, the preequilibrium process and the direct reaction mechanism. Among the available nuclear reaction codes, the TALYS simulation package (v. 1.9) [8] has been considered in this work for the theoretical analyses of the cross sections for the $^{nat}V(p, x)^{47}Sc$ reaction. The code is very versatile and a variety of different models are implemented in for the description of the processes in the interaction between a projectile and a target, going from phenomenological models to modern microscopic approaches. In particular, inside the code, there are four models for the preequilibrium emission and six models for the description of the nuclear level density, resulting therefore in several combinations of models that can be considered for any reaction. Different possibilities can be examined but in literature many calculations refer to the use of two specific combinations of models: the so-called "TALYS default" (based on the default options of the code) and "TALYS adjusted" [10] (based on another configuration usually used to replace the default one). None of the options is in some cases either the best choice or the unique possibility and therefore alternatives can be also investigated. As described in Ref. [11], a possibility is given by considering all the combinations of models and using descriptive statistical quantities, mainly quartiles. With this approach, instead of plotting all the curves, only a statistical band is constructed from the interquartile range, given by the difference between the third (Q3) and the first (Q1) quartiles. Starting from the band, for each energy a theoretical reference curve (BTE which stands for "Best Theoretical

Evaluation") can be introduced as the centroid of the band, to which an uncertainty (given by the half-width of the band) is associated:

$$\sigma_{BTE} = \frac{Q_1 + Q_3}{2} \qquad \Delta_{BTE} = \frac{Q_3 - Q_1}{2}. \tag{1}$$

The new curve together with the corresponding uncertainty can be used for the further estimations of yields and purities of the produced nuclides. Nevertheless, another possible choice can be represented by the improvements on a specific model, which is the approach used in this work to better reproduce the trend of the experimental data. The used procedure is not a pure theoretical approach but more likely can be described as a phenomenological one, trying to use already developed models and adjust some of their parameters to obtain new curves to reproduce the data, in view of the practical applications of the produced radionuclides. In particular the attention has been addressed to the three microscopic models for the level densities, which are based on Hartree-Fock Methods (HFM), comprehensive of HF plus Bardeen-Cooper-Schrieffer (HF-BCS) [12] and HF Bogoliubov (HFB) [13, 14]. Inside TALYS the HFM level densities are tabulated for different energy values and subsequently are rescaled according to the following transformation:

$$\rho(E, J, \pi) = exp^{(c\sqrt{E - p})} \rho_{HFM}(E - p, J, \pi) \tag{2}$$

which depends on the two parameters c and p. The former acts as an overall normalization while the latter could represent an energy shift caused by pairing or even shell effects. Inside the code each microscopic model has tables of parameters that allow to obtain new level densities starting from the pure theoretical HFM ones. Nevertheless, the curves of the microscopic models, with their tabulated values of the parameters, do not always describe well the cross sections of a given reaction, but the possibility to vary the parameters is given, as stated in the TALYS manual [15] as well as in the RIPL library [16].

In this work a combination of models has been considered: for the nucleon-nucleus interaction the so-called JLM (Jeukenne-Lejeune-Mahaux) semi-microscopic optical potential model [17, 18], while for the preequilibrium process the default option, based on the exciton model with transition rates calculated numerically. Regarding the nuclear level density, a tuning of the two parameters has been performed considering the model based on HF-BCS method that is denoted inside the code with the keyword *ldmodel* 4, in order to obtain a better agreement between the theoretical curves and the experimental data for all the nuclides studied in the PASTA project. Specifically, an initial grid search has been performed on the two parameters considering all the nuclides involved and an initial set of values has been obtained by a qualitative agreement, based on a visual analysis, between the TALYS curves and the data. As a second step a MINUIT [19] global chi-square minimization procedure, using the combination of gradient and simplex optimizations, has been implemented to refine the solutions, arriving to the final optimal set used to obtain new theoretical curves which are referred to as "TALYS modified". The final set reported in Tab. 1 is not intended as the unique possible solution, since other possibilities could also be explored with different parameters. Nevertheless, with the found solution a very high accuracy in the description of the cross sections has been achieved, especially for the main contaminant.

Table 1. Final set of values of the level density parameters.

	^{42}K	^{43}K	^{44}Sc	^{46}Sc	^{48}Sc	^{48}V	^{48}Cr
c	1.27443	1.09559	1.97876	0.339592	-	-	0.234421
p	-	0.140468	1.11855	-1.17731	0.449064	-0.25841	-

3 Results and discussion

The reaction with protons on natural vanadium targets has been considered, within the energy range 26-70 MeV, for the production of ^{47}Sc, ^{43}Sc, ^{44g}Sc, ^{44m}Sc, ^{46}Sc, ^{48}Sc, ^{42}K, ^{43}K, ^{48}V, ^{48}Cr, ^{49}Cr, ^{51}Cr. A complete discussion of all the nuclides studied in the PASTA project is given in [9], while in this paper only the ^{47}Sc and ^{46}Sc cases are shown with BTE and TALYS modified approaches. In the following figures the comparison between the theoretical cross

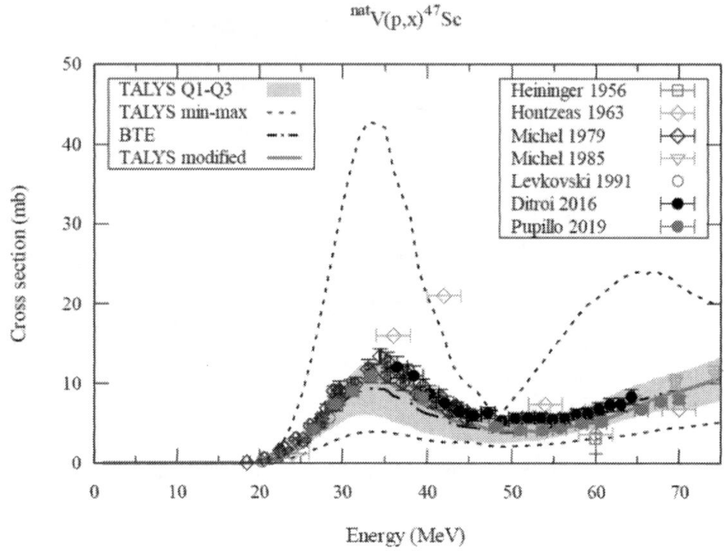

Fig. 1. Cross section for the ^{nat}V(p,x)^{47}Sc with the BTE approach and with TALYS modified.

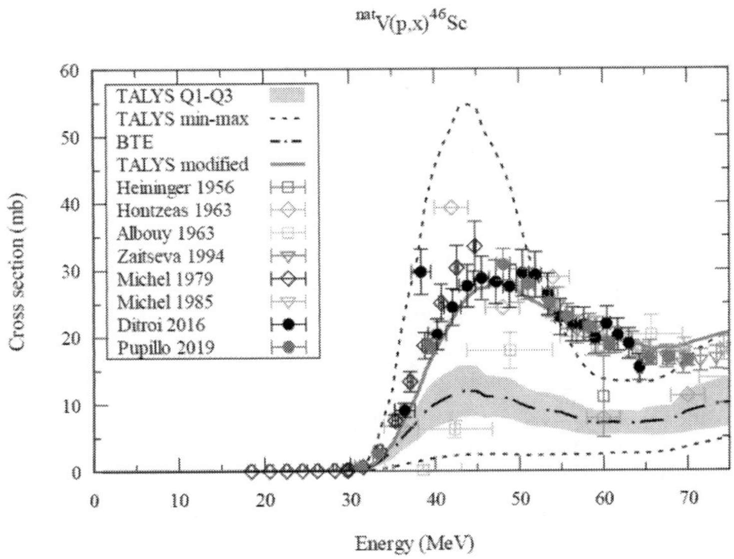

Fig. 2. Cross section for the $^{nat}V(p,x)^{46}Sc$ with the BTE approach and with TALYS modified.

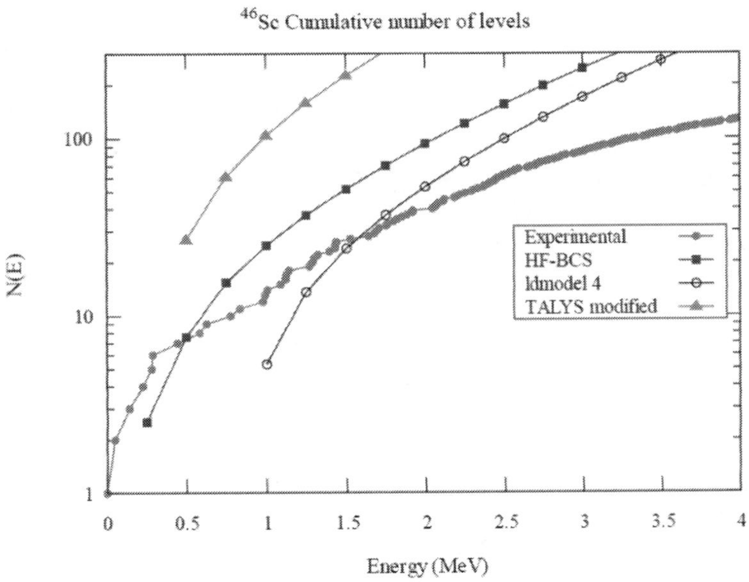

Fig. 3. Comparison among theoretical and experimental cumulatives.

sections and the experimental data [7, 20-27] and the comparison among the cumulatives (only for ^{46}Sc) are shown.

In Figs. 1 and 2 all the models are represented with the statistical representation described in the previous section. The dashed lines correspond to the minimum and maximum among the combinations, the shaded area is the interquartile range, containing the 50% of the models, while the dash-dot line is the BTE curve (the central curve of the band). In Fig. 1 the trend of the recent data is quite well described by the BTE curve, while in Fig. 2 the statistical approach allows to observe that 50% of the models does not reproduce the trend of the recent data not only in the region around the cross section peak but also in the area of high energy (60-70 MeV), where the TALYS models show a different behavior if compared to the one of the data. Instead of the BTE curve, a better description is given by the solid line, TALYS modified, which is the curve obtained with the tuning procedure of the nuclear level density parameters. In this case the curves for both ^{47}Sc and ^{46}Sc well reproduce the trend of the recent data in all the energy regions resulting in more accurate estimations of their production yields and therefore also of the radionuclidic purity of the nuclide of interest.

In Fig. 3 the comparison among the experimental cumulative, represented with points, and three theoretical cumulatives is shown, as an example, only for ^{46}Sc. The one represented with squares corresponds to the curve obtained with the HF-BCS method (whose values are tabulated inside the code) for which both the *c* and *p* parameters are equal to zero. The cumulative with open circles is the one of the model indicated with *ldmodel 4* which is derived from the HF-BCS cumulative using the tables of the parameters inside the code. The last cumulative with triangles is obtained with the new values of *c* and *p* found in this work. From the figure a disagreement between the TALYS modified curve with the experimental one is evident, but also the other two theoretical cumulatives do not provide an accurate description. This behavior could be explained by considering that the models, available at the moment, do not take into account all the possible phenomenological effects and significant contributions, as the isospin dependence in statistical processes (as already stated in the literature [28]). This may underline the need of the optimizations to achieve a better description of the cross sections, that was the main goal of this work, in order to evaluate the production yields and the contribution to the purity of ^{47}Sc in a more accurate way, as stated in [9]. Therefore, to show the impact of this optimization procedure of the cross sections on the applications in nuclear medicine, another step is required: the evaluation of the production yields of ^{47}Sc and ^{46}Sc, which have been calculated using the new curves and considering the same irradiation conditions identified in [7]. An energy range of the beam of 19-30 MeV, corresponding to a 1.21 mm thick target, has been used in the calculations to maximize the production of ^{47}Sc while minimizing the co-production of ^{46}Sc. In Tab. 2 the estimations of the thick-target yields (TTY) at the end-of-bombardment (EOB) are reported for two irradiation plans of 24 h and 80 h for TALYS modified for the two nuclides in comparison with the values obtained with ISOTOPIA (an online tool developed and maintained by IAEA), which instead seems to underestimate their production, especially for the main contaminant.

Table 2: TTY at EOB for ^{47}Sc and ^{46}Sc for two irradiation time comparing the results obtained with the TALYS modified curve and the ones from ISOTOPIA.

	TALYS modified		ISOTOPIA	
Irradiation time	^{47}Sc	^{46}Sc	^{47}Sc	^{46}Sc
24 h	33.48 MBq/μA	0.0137 MBq /μA	25.29 MBq /μA	0.0013 MBq /μA
80 h	89.24 MBq /μA	0.0453 MBq /μA	67.43 MBq /μA	0.0042 MBq /μA

Subsequent steps are represented by the calculation of the radionuclidic purity (RNP) as shown in [6, 29], and dosimetric calculations to assess the dose imparted to the patients.

4 Conclusions

In this work an improvement of the theoretical cross sections for the proton-induced reaction on natural vanadium has been presented working on the nuclear level density parameters of the model based on HF-BCS method implemented inside the TALYS code. In particular the obtained improvements for the case of ^{47}Sc and for its main contaminant, ^{46}Sc, have been shown. The new cross section curve reproduces better the trend of the experimental data, resulting in more accurate and reliable estimations of the production yields to be used for subsequent dosimetric studies, in view of the applications of the theranostic radionuclide.

The tuning procedure is completely general and can also be applied in other studies concerning the production of nuclear medicine radionuclides, as indicated in [6] for the production of ^{47}Sc with proton beams on titanium targets or also in [29] for the production of ^{52}Mn for diagnostic purposes.

References

1. A. R. Jalilian et al., Curr. Radiopharm., **14**, 306-314 (2021), http://dx.doi.org/10.2174/1874471013999200928162322
2. C. Müller et al., Br. J. Radiol., **91**, 20180074 (2018), https://doi.org/10.1259/bjr.20180074
3. National Nuclear Data Center (NNDC), NuDat 2.8 database, https://www.nndc.bnl.gov/, (accessed October 2021)
4. S. Loveless et al., EJNMMI Res, **9**, 42 (2019), https://doi.org/10.1186/s13550-0190515-8
5. S. M. Qaim, Medical radionuclide production (De Gruyter 2019) https://doi.org/10.1515/9783110604375
6. F. Barbaro et al., Theoretical study of ^{47}Sc production for theranostic applications using proton beams on enriched titanium targets, (communication at this conference)
7. G. Pupillo et al., J. Radioanal. Nucl. Chem., **322**, 1711-1718 (2019), https://doi.org/10.1007/s10967-019-06844-8
8. S. Goriely et al., Astron. Astrophys., **487**, 767-774 (2008), https://doi.org/10.1051/0004-6361:20078825
9. F. Barbaro et al., Phys. Rev. C, **104**, 04461 (2021), https://doi.org/10.1103/PhysRevC.104.044619
10. C. Duchemin et al., Phys. Med. Biol., **60**, 931 (2015), https://doi.org/10.1088/00319155/60/3/931
11. A. Colombi et al., Nucl. Technol. (2021), https://doi.org/10.1080/00295450.2021.1947122
12. F. Minato, J. Nucl. Sci. Technol., **48**, 984-992 (2011), https://doi.org/10.1080/18811248.2011.9711785
13. S. Hilaire and S. Goriely, Nucl. Phys. A, **779**, 63-81 (2006), https://doi.org/10.1016/j.nuclphysa.2006.08.014
14. S. Hilaire et al., Phys. Rev. C, **86**, 064317 (2012), https://doi.org/10.1103/PhysRevC.86.064317
15. A. Koning et al., User Manual of Talys-1.9, (2017), https://www.talys.eu/

16. R. Capote et al., Nucl. Data Sheets, **110**, 3107-3214 (2009), https://doi.org/10.1016/j.nds.2009.10.004
17. J. P. Jeukenne et al., Phys. Rev. C, **15**, 10 (1977), https://doi.org/10.1103/PhysRevC.15.10
18. J. P. Jeukenne et al., Phys. Rev. C, **16**, 80 (1977), https://doi.org/10.1103/PhysRevC.16.80
19. F. James and M. Roos, Comput. Phys. Commun., **10**, 343-367 (1975), https://doi.org/10.1016/0010-4655(75)90039-9
20. C. Heininger and E. O. Wiig, Phys. Rev., **101**, 1074-1076 (1956), https://doi.org/10.1103/PhysRev.101.1074
21. S. Hontzeas and L. Yaffe, Can. J. Chem., **41**, 2194-2209 (1963), https://doi.org/10.1139/v63-320
22. G. Albouy et al., J. Phys., **24**, 67-68 (1963), https://doi.org/10.1051/jphysrad:0196300240106700
23. R. Michel et al., Nucl. Phys. A, **322**, 40-60 (1979), https://doi.org/10.1016/03759474(79)90332-4
24. R. Michel et al., Nucl. Phys. A, **441**, 617-639 (1985), https://doi.org/10.1016/03759474(85)90441-5
25. V. N. Levkovskij, (Inter-Vesi: Moscow, Russia, 1991)
26. N. G. Zaitseva et al., Radiochim. Acta, **65**, 157-160 (1994), https://doi.org/10.1524/ract.1994.65.3.157
27. F. Ditrói et al., Nucl. Instrum. Methods Phys. Res. B, **381**, 16-28 (2016), https://doi.org/10.1016/j.nimb.2016.05.015
28. S. M. Grimes et al., Phys. Rev. C., **5**, 85 (1972), https://doi.org/10.1103/PhysRevC.5.85
29. M. P. Carante et al., A new route to produce ^{52g}Mn with high purity for MultiModal Imaging, (communication at this conference)

AUTHOR INDEX

AUTHOR INDEX

EDP Sciences – Web of Conferences
17, Avenue du Hoggar
Parc d'Activité de Courtabouf
BP 112
F-91944 Les Ulis Cedex A
France

ISBN 978-1-7138-4832-5

CURRAN ASSOCIATES INC.
proceedings.com

10th International Conference on Environment and Renewable Energy (ICERE 2024)

E3S Web of Conferences Volume 514 (2024)

Nha Trang, Vietnam
25 – 27 February 2024

edp sciences